T0319262

WIRE TECHNOLOGY

WIRE TECHNOLOGY
PROCESS ENGINEERING AND METALLURGY

ROGER N. WRIGHT

Amsterdam • Boston • Heidelberg • London • New York
Oxford • Paris • San Diego • San Francisco • Singapore
Sydney • Tokyo

ELSEVIER   Butterworth-Heinemann is an imprint of Elsevier

Butterworth-Heinemann is an imprint of Elsevier
30 Corporate Drive, Suite 400, Burlington, MA 01803, USA
Linacre House, Jordan Hill, Oxford OX2 8DP, UK

Library of Congress Cataloging-in-Publication Data
Wright, Roger N.
Wire technology : process engineering and metallurgy / Roger N. Wright.
   p.   cm.
ISBN 978-0-12-382092-1
1. Wiredrawing. I. Title.
TS270.W75   2011
671.8'42–dc22

2010025887

British Library Cataloguing in Publication Data
A catalogue record for this book is available from the British Library.

ISBN: 978-0-12-382092-1

For information on all Butterworth-Heinemann publications
visit our Web site at www.books.elsevier.com

Printed in the United States of America
11 12   11 10 9 8 7 6 5 4 3 2 1

DEDICATION

To my wife, Patricia, who has learned that every room is an office, and every table is a desk.

CONTENTS

PREFACE

Being in a family with several generations of professional practitioners in metals processing and the teaching thereof, I suppose my writing of this book was inevitable. Even so, I must clearly acknowledge two strong influences outside of the family sphere. The first was the late Walter A. (Al) Backofen, professor of metallurgy and materials science at MIT a half-century or so ago. While I received the benefit of some of his lectures, his major impact was by way of his book *Deformation Processing*, Addison-Wesley, 1972. This book was the first that I am aware of to teach deformation processing with major emphasis on Δ, the shape of the deformation zone. To be sure, Δ or its equivalent was utilized in some of the more enlightened mid-twentieth century wire drawing research (most notably that of J. G. Wistreich) and citations of the importance of deformation zone geometry can be found in the literature of the 1920's. However, Backofen powerfully employed it as a teaching tool, bringing together a considerable array of mechanical analyses, process designs and mechanical metallurgical phenomenology. As a young metallurgist, I assumed that just about everybody used Δ, only to find out that its work-a-day industrial applications had been minimal. In this context, I applied it (arguably even over applied it) every chance that I had, and in the wire industry I believe it has been of significant value. In any case, it is central to much of this book, and I have Professor Backofen to thank.

The other influence that I would like to cite was Dr. Alan T. Male, my manager during the years that I spent at Westinghouse Research Laboratories. Alan was, of course, renown for his development of the ring compression test that quantifies friction in forging (a brilliant application of deformation zone geometry, incidentally). Moreover, he had been a faculty member at The University of Birmingham and had an instinctive and synergistic approach to applying rigorous research technique and perspective to industrial processing systems. He, early on, directed my involvement in a wide variety of sophisticated wire processing studies, as well as in the supervision of industrial society seminars and short courses. When I left Westinghouse to join the faculty at Rensselaer Polytechnic Institute in 1974, I had been given a thorough education in wire processing, to go with my broader backgrounds in metallurgy and metals processing.

Addressing the subject at hand, I have written this book in the style of an upper level undergraduate, or possibly graduate level text, acknowledging that one is not likely to find such a course on wire processing, except perhaps in Eastern Europe. This approach has allowed me to use directly much of my experience in technology-focused short courses, as well as my experience in teaching undergraduates and graduate students at Rensselaer. I have written it with the hope that it will be useful for self study and continuing education offerings, as well as serving as a desk reference. At this point in time, I believe that it occupies a unique position in the engineering literature.

Finally, I would like to thank the most helpful staff at Elsevier, Inc., particularly Haley Salter and Kiru Palani, for their patient handling of this, my first, book.

Roger N. Wright, ScD, PE, FASM, FSME

ABOUT THE AUTHOR

Roger N. Wright, professor of materials engineering at Rensselaer Polytechnic Institute, has contributed broadly to the literature in the areas of metallurgy and metals processing, and is active as a short-course lecturer and consultant. Prior to joining Rensselaer, he was a senior staff member at Westinghouse Research Laboratories and at Allegheny Ludlum Steel Corporation. He holds B.S. and Sc.D. degrees in metallurgy from Massachusetts Institute of Technology. He is a registered professional engineer and a fellow of ASM International and of the Society of Manufacturing Engineers.

CHAPTER ONE

The General Idea

Contents

 ## 1.1. CONCEPTS

1.1.1 Drawing

The concept of drawing addressed in this book involves pulling wire, rod, or bar through a die, or converging channel to decrease cross-sectional area and increase length. In the majority of cases the cross section is circular, although non-circular cross sections may be drawn and/or created by drawing. In comparison to rolling, drawing offers much better dimensional control, lower capital equipment cost, and extension to small cross sections. In comparison to extrusion, drawing offers continuous processing, lower capital equipment cost, and extension to small cross sections.

1.1.2 Wire, rod, and bar

In general, the analyses of wire, rod, and bar drawing are similar, and we may use the term **workpiece**, or simply the term *"wire,"* when there is no distinction to be drawn. However, there are major practical and commercial issues to be addressed among these terms. **Bar** drawing usually involves stock that is too large in cross section to be coiled, and hence must be drawn straight. Round bar stock may be 1 to 10 cm in diameter or even larger. Prior to drawing, bar stock may have been cast, rolled, extruded, or swaged

Wire Technology
ISBN 978-0-12-382092-1, DOI: 10.1016/B978-0-12-382092-1.00001-4

(rotary cold forged). **Rod** drawing involves stock that may be coiled, and hence may be delivered to the die from a coil, and taken up as a coil, on a block or capstan. Round rod stock will often have a 0.3 to 1 cm diameter, and will often have been cast and/or rolled prior to drawing. **Wire** drawing involves stock that can be easily coiled and subjected to sequential or tandem drawing operations with as many as a dozen or more draws occurring with a given drawing machine. Each drawing operation or "pass" will involve delivery of the wire to the die from a coil on a capstan, passage through the die, and take-up on a capstan that pulls the wire through the die. **Fine wire** drawing typically refers to round wire with a diameter of less than 0.1 mm, and **ultra-fine wire** drawing typically refers to round wire as fine as 0.01 mm in diameter.

1.1.3 Materials

Essentially any reasonably deformable material can be drawn, and the general analysis is the same regardless of the wire, rod, or bar material. The individual technologies for the major commercial materials, however, involve many nuances. The drawing technologies are often divided into **ferrous** (steel) and **non-ferrous** and **electrical** (usually copper and aluminum), although there is specialty production and research and development interest in such high-value-added products as thermocouple wire, precious metal wire, biomedical wire, wire for high temperature service, superconducting wire, and so on.

Apart from the material drawn, drawing technology depends substantially on the materials used for **dies** ("carbide," diamond, tool steel) and on the materials or formulations used for **lubricants** and coatings.

1.2. HOW DOES DRAWING WORK?

1.2.1 Why not simply stretch the wire, rod, or bar?

It can be argued, at least in principle, that some of the objectives of drawing could be achieved by simply stretching the wire with a **pulling force**. The cross section could be reduced and elongation accomplished, but dies would not be needed and the friction and metal flow issues presented by the die could be avoided.

The principal problem with just stretching the wire with a pulling force is the necking phenomenon. Basically, after a certain amount of uniform stretching, all further elongation will be concentrated at a single location (a neck), which will rapidly thin and break. This occurs because the decrease in cross-sectional area eventually weakens the wire more than any strengthening that occurs by work hardening. Heavily drawn wire will have little or no work-hardening capability, and will neck almost at once if subjected to simple stretching. Although some complex "dieless" drawing systems have been invented, simple stretching has only limited application because of its vulnerability to necking.

1.2.2 A simple explanation of the drawing process

In the drawing process, a **pulling force** and a **pressure force**, from the die, **combine** to cause the wire to extend and reduce in cross-sectional area, while passing through the die, as schematically illustrated in Figure 1.1. Because of this combined effect, the pulling force or **drawing force** can be less than the force that would cause the wire to stretch, neck, and break downstream from the die. On the other hand, if a reduction too large in cross-sectional area is attempted at the die, the drawing force may break the wire. In commercial practice, engineered pulling loads are rarely above 60% of the as-drawn strength, and the area reduction in a single drawing pass is rarely above 30 or 35%, and is often much lower. A particularly common reduction in non-ferrous drawing is the **American Wire Gage (AWG)** number, or about 20.7%. Many drawing passes are needed to achieve large overall reductions.

1.2.3 Comparison to other processes

The use of pulling or pushing forces, together with dies or rolls, is common to many deformation processes[1,2], as shown in Figure 1.2. Figure 1.2a illustrates

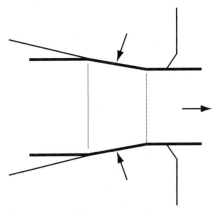

Figure 1.1 Schematic illustration of forces in drawing.

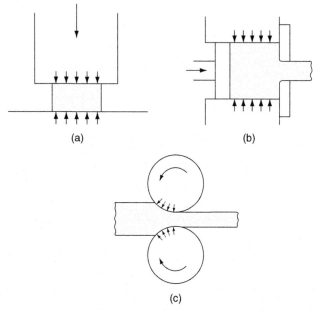

Figure 1.2 Schematic illustration of (a) forces in forging or upsetting, (b) some of the forces in extrusion, and (c) roll motion and roll force in rolling. From G.E. Dieter, *Mechanical Metallurgy*, Third Edition, McGraw-Hill, Boston, MA, 1986, p. 504. Copyright held by McGraw-Hill Education, New York, USA.

the basics of a simple forging or upsetting operation, and Figure 1.2b and c illustrate extrusion and rolling operations, respectively. Many other variations exist. For example, rod or strip can be reduced by pulling through undriven rolls, and so on.

The term "drawing" is used to describe a number of metallurgical processing operations, and when searching titles in the metalworking or intellectual property literature, be careful not to confuse references to *deep* drawing of sheet metal, drawing aspects of forging, or steel tempering operations referred to as drawing, and so on, with the pulling operations outlined in this book.

1.2.4 Overall process hardware

In addition to the die, held in a **die block**, a basic drawing operation involves a **payoff** and a **take–up**, as illustrated in Figure 1.3. Also necessary is a system for applying lubricant to the wire before it enters the die. Figure 1.3 schematically illustrates a **soap box**, which contains a solid powdered-soap lubricant that the wire is pulled through prior to die entry. With liquid

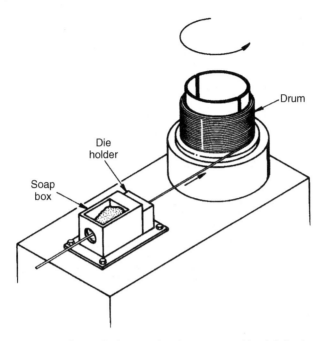

Figure 1.3 Illustration of a single die wire drawing system with a lubrication application box, a die (in a die holder), and a rotating drum to apply tension and take up the wire. From B. Avitzur, *Handbook of Metal-Forming Processes*, John Wiley & Sons, New York, 1985, p. 195. Copyright held by B. Avitzur, Allentown, PA, USA.

lubrication, the lubricant may be directed in a stream at the die entry, and the drawing system may even be submerged in lubricant. Figure 1.3 shows the case of a single die system. As discussed in sections 3.3 and 3.4, drawing systems often employ successive or tandem dies and pulling operations.

A drawing operation must have a method for **pointing** the wire. Pointing involves reducing the "front" end diameter of the wire sufficiently to allow it to be initially passed through the die and gripped en route to initial winding onto the take–up.

1.3. QUESTIONS AND PROBLEMS

1.3.1 One of the processes schematically illustrated in Figure 1.2 is particularly well suited to very long workpiece lengths, as is drawing. Which process is this? Why are the other two illustrated processes not as well suited?
Answers: Rolling is particularly well suited to very long workpiece lengths, such as coils, because it is a continuous process. Forging involves a limited

workpiece, which is constantly changing shape. Extrusion usually involves a
limited workpiece, as well, although some "continuous" extrusion technologies have been developed involving billet-to-billet juxtapositions or frictional billet pressurization with belt or chain systems.

1.3.2 List some ways that wire, rod, and bar can be pointed. Do not be
afraid to use your imagination.

Answers: These ways include rotary swaging (see Section 18.6.3), rolling,
machining, stretching, and chemical attack.

1.3.3 Why is cross-sectional dimensional control much better in drawing
than in rolling?

Answer: The die is one piece in drawing with wear the only common
source of cross-sectional dimension change. Rolling forces cause changes in
the roll gap, and bar rolling involves complex shape changes.

1.3.4 Wire breakage during drawing can significantly impact the profit-
ability of a production facility. Cite at least two costly aspects of a wire
break.

Answers: Production time is lost restringing the machine; wire lengths too
short for continued drawing may have to be scrapped; and wire breakage
may indicate that large numbers of flaws are generated, implying possible
rejection of the drawn-wire product, and mandating increased quality
control and process troubleshooting.

CHAPTER TWO

A Brief History of Technology

Contents

2.1. ANCIENT AND EARLY TECHNOLOGY

Rod and wire technologies are of ancient origin, although some distinction must be made between wire making and wire drawing. Gold wire was incorporated into the adornments of the pharaohs by Egyptians as early as 3000 BC, and technique development probably predates this era. It is likely that the ancients cut strips from hammered foil and then drew folded strips though stone dies as the initial step in wire making. Cross-sectional consistencies indicate that drawing dies were available to such craftsmen. It is thought that holes were bored in natural stone with the aid of pointed sticks and sand/tallow abrasive media.

There are interesting references to wire in very early literature, particularly in Homer's *Odyssey* (*The Songs of the Harper*) and in the <u>Old Testament</u> (Exodus 28:14 and 39:3). By fifth century BC, the Persians were drawing 0.55 mm bronze wire with iron draw plates, implying that they may have understood the concepts of multiple passes and interpass annealing. Interesting references to drawing technology were made by the Roman tribune Claudius Claudianus toward the end of the Roman empire in 400 AD.

Moving ahead to the Middle Ages, the monk Theophilus Presbyter wrote about drawing technology around 1125, and it is clear that commercial practices were emerging. A document written in Paris around 1270 notes that:

1. The wire drawer must thoroughly understand his trade and have sufficient capital at his command.
2. The wire drawer may have as many apprentices and servants as he wishes, and may work nights as much as he pleases.

Wire Technology
ISBN 978-0-12-382092-1, DOI: 10.1016/B978-0-12-382092-1.00002-6

3. The wire drawer need pay no taxes on anything relating to his trade which he buys or sells in Paris.

4. Apprentices to wire drawers will serve ten years without pay and then be paid a premium of 20 sous.

Nuremberg was apparently a major center for Middle Ages and Renaissance wire technology, with documentation from the fourteenth to middle sixteenth centuries found in the *Hausbuch der Mendelschen Zwölfbruderstiftung zu Nürnberg*. Major developments are attributed to Rudolph von Nuremberg. In the early fourteenth century he utilized water power and camshaft-driven draw benches. Previous to this, the only practical sources of power were manual, which involved such expedients as hand lever devices called "brakes" and swinging body motion utilized by harnessed "girdlemen." The rather effective dies prepared from hard stone by the Egyptians were followed in later millennia by easily worked, but rapidly wearing iron and steel plates. An illustration of a swing-assisted medieval rod drawer with tongs and drawing plate is shown in Figure 2.1.

Figure 2.1 Illustration of medieval wire drawing, as presented in the *Hausbuch der Mendelschen Zwölfbruderstiftung zu Nürnberg*.

Figure 2.2 Albrecht Durer's 1489 painting, *The Wire Drawing Mill*. (Staatliche Museen, Berlin)

The development of lubricants has been a vital, if subtle, aspect of the history of drawing. The earliest drawing is thought to have depended on animal fat or tallow. This was augmented by particulate matter in the form of lime, carbon black, tars, powdered coal, and graphite. Reactive lubricant additions that maintained lubricant integrity at elevated temperatures were later introduced. Soft metal coatings were implemented in some cases. A particularly intriguing development was Johan Gerdes' discovery of the "sull coat" (actually thin iron oxide) around 1632. He allegedly utilized the superior lubrication response of wire exposed to human urine. Aspects of Gerdes' discovery were employed for the next two centuries.

The German artist Albrecht Durer painted *The Wire Drawing Mill* in 1489 with an apparent water power source, as shown in Figure 2.2. By the fifteenth century none other than Leonardo da Vinci was sketching drawing blocks and noted that: "Without experience you can never tell the real strength with which the drawn iron resists the drawing plate."

2.2. THE NINETEENTH CENTURY

The "industrial revolution" started at the end of the eighteenth century, and the nineteenth century involved rapid improvements in wire technology, particularly in regard to productivity. Beginning in Portsmouth,

England, in 1783 with Henry Cort's implementation of grooved rolls through the evolution of Belgian looping mills in 1860, and George Bedson's continuous rod rolling mill installed in 1862 at the Bradford Ironworks in England, rod rolling developments allowed and necessitated the processing of very long lengths of rod and wire. In this context, the first continuous drawing machines appeared in Germany and England around 1871.

Prior to the nineteenth century, wire production was motivated by the demands of the decorative arts, the military, and the textile industry (card wire). Much of nineteenth century progress was interrelated with the rapid growth of new product markets. The following products and the dates of their inception are noteworthy: wire rope (1820), telegraph wire (1844–1854), wire nails (1851–1875), bale ties and barbed wire (1868), telephone wire (1876), screw stock (1879), coiled wire springs (1879), and woven wire fence (1884). Also important were large, but unstable, markets for women's apparel items such as hoop skirts (crinoline wire), corsets, and hairpins.

Development of cast iron and tool steel dies was undertaken in conjunction with the increased productivity of the nineteenth century, and natural diamonds were employed for sizes below one millimeter.

2.3. THE TWENTIETH CENTURY

Twentieth century wire processing advances included such items as in-line annealing and heat treatment, sophisticated wire-handling systems that allowed high drawing speeds, multiple strand drawing systems, and a variety of process automation and control innovations. The engineering of drawing systems was helped greatly by a number of practical results from research and theoretical analysis. Particularly noteworthy were the published efforts of Körber and Eichinger,[3] Siebel,[4,5] Sachs,[6,7] Pomp,[8] Wistreich,[9] and Avitzur.[10,11]

However, the most significant twentieth century advances have been in the area of die materials. Vastly improved die performance/cost ratios were enabled by the development of cemented carbide and synthetic diamonds. The cemented carbide development is generally credited to two independent German investigators, Baumhauer and Schröter, who incorporated cobalt and tungsten carbide powders into a sinterable compact in 1923. The product was developed commercially by the firm of Friedrich Krupp under the trade name of Widia. This economical and highly wear-resistant material quickly supplanted most die materials, even threatening to displace diamond dies.

The use of natural diamond dies for fine gages persisted; however, and natural diamond dies and modern carbides were joined in 1974 by synthetic diamond dies, first introduced by the General Electric Company under the name Compax. This product, and subsequent variations and competing products, utilized synthetic diamond powder first developed by General Electric in 1954.

Twentieth century lubrication developments involved the use of a number of chemically engineered soaps, gels, and emulsions, including synthetic as well as natural products. Major attention was devoted to lubricant removal and disposal as well as to environmental impact.

 ## 2.4. FURTHER READING

The remarks in the previous section are a short summary abstracted from a number of more extensive publications. For more information, the interested reader is directed to a number of reviews, which, in turn, reference historical sources.[12–15]

 ## 2.5. QUESTIONS AND PROBLEMS

2.5.1 Read the technology references in Exodus 28:14 and 39:3, preferably in more than one translation of the Bible. What sorts of equipment or manual skills are implied?

Answers: The New Revised Standard Version (NRSV) of the Holy Bible refers to "two chains of pure gold, twisted like cord" in Exodus 28:14, whereas the King James Version (KJV) refers to "two chains *of* pure gold at the ends; *of* wreathen work." In Exodus 39.3, it is noted in the NRSV that "gold leaf was hammered out and cut into threads," whereas the KJV says "And they did beat the gold into thin plates, and cut *it into* wires." It seems that drawing is not described, but rather the cutting of strips from hammered foil. Cutting tools would have been required. One wonders if the chains referred to were cut whole from plate or involved wire joined into loops.

2.5.2 Compare one or more of the Paris regulations created around 1270 with practices in today's wire industry.

Answers: The need for capital remains an issue, to say the least (regulation 1). Clearly the city of Paris was offering incentives for a resident wire industry, as regions seeking to attract industry still do(regulation 3). The practices of regulations 2 and 4 are not as common today.

2.5.3 Examine Figure 2.1 carefully. What drawing speed and production rate do you think the craftsman is capable of?

Answer: Routine hand labor is generally at speeds of 1 m/s, and this would be a good guess for the worker in Figure 2.1. The rod appears to have a diameter of roughly 2 cm. Thus the volume drawn in 1 s would be near 300 cm^3, and the volume for an hour of actual drawing would be somewhat over 1 m^3. If the product were iron base, the mass for an hour of actual drawing would be under 10,000 kg or a rate of roughly ten tons per hour. This does not factor in the down time between pulls and time for rest. The drawer would probably be doing well to draw a ton or two per hour.

2.5.4 The development of the American "heartland" involved numerous expanded markets for wire, and there is even a pertinent citation at the Alamo in San Antonio, Texas. Similar observations can be made for central Europe. Cited in Section 2.2 are examples of wire products and applications such as telegraph and telephone wire, bale ties and barbed wire, and woven fence wire. Moreover, the ubiquitous availability of wire led to many secondary products often made at home or by traveling "tinkers." Think of some possible home implements that could have been made of wire.

Answers: The interested reader is directed to *Everyday Things Wire* by Slesin et al.[16] Examples of cages, traps, baskets, wine caddies, condiment sets, grills, toasters, bottle carriers, egg holders, and platters are shown, as well as illustrations of whisks, beaters, whips, griddles, forks, mashers, strainers, hangers, light and lamp protectors, weeders, pickers, and endless toys and "gifts."

CHAPTER THREE

Twentieth Century Equipment Concepts

Contents

3.1. OVERVIEW

As stated in Section 1.1.2, the terms bar, rod, and wire often imply a certain mode of processing, or process flexibility, especially regarding the ability to coil the product during process sequences. The related drawing equipment can be roughly categorized as *benches*, *blocks*, and *multiple-die machines*. Countless variations and subtleties exist regarding these equipment types, and a comprehensive treatment of wire drawing machinery is beyond the scope of this text. However, some useful simplifications and characteristics are shown in this chapter.

3.2. BENCHES

While the term "bench" has been applied to a variety of wire processing assemblies, this text will regard drawing benches as involving the simple pulling of straight lengths, where, in the simplest cases, the length achievable is limited by the length of the bench. It should be noted, however, that continuous bench-type machines have been developed, such as systems applying a "hand-over-hand" pulling technique. In any case, simple bench drawing does not generally involve coiling of the drawn workpiece, although benches, particularly continuous benches, are often in tandem with straightening and cutting machines.

Wire Technology
ISBN 978-0-12-382092-1, DOI: 10.1016/B978-0-12-382092-1.00003-8

In addition to use with uncoilable workpieces (heavy-gage stock, bend-sensitive stock, etc.), benches are useful for certain laboratory or development studies and for short lengths of specialty items. Drawing bench speeds do not generally exceed 100 m/min.

3.3. BLOCKS

When bar or rod is sufficiently robust or of small enough diameter to permit coiling, block drawing may be employed. The block involves a **capstan** or **bull block** to which the rod is attached. The powered bull block turns, pulling the rod through the die and coiling the as-drawn rod on the bull block. **Single block** or capstan drawing is often undertaken, although **multiple-block** systems are common, with the rod wrapped a few times around each capstan before entering the next, smaller gage die. The capstans transmit pulling force to the rod by way of the frictional contact of the rod wraps on the capstan surface. A schematic illustration of block drawing was given in Figure 1.3. Figure 3.1 shows a commercial drawing machine with the capstans and die stations clearly indicated. Block drawing speeds are often in the range of 100 to 200 m/min, with the **drawing speed, v**, as:

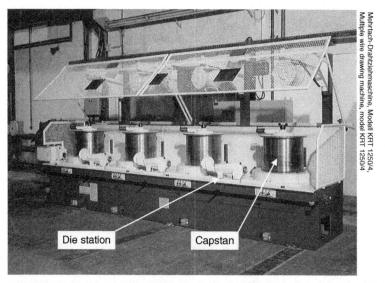

Mehrtach-Drahtziehmaschine, Modell KRT 1250/4,
Multiple wire drawing machine, model KRT 1250/4

Die station   Capstan

Figure 3.1 Commercial multiple-block drawing machine with capstans and die stations indicated. (Courtesy of Morgan-Koch Corporation)

$$v = \pi\, D\, \omega, \qquad\qquad (3.1)$$

where D is the block diameter and ω is the block speed in revolutions per unit time. Higher speed multiple-block systems are discussed in Section 3.4.

3.4. MULTIPLE-DIE MACHINES

As the rod or wire gets smaller in diameter, high-speed, **multiple-die machines** become practical and necessary for commercial productivity. These may be of the multiple or tandem capstan variety or may involve a single, multiple-diameter capstan of the **"stepped cone"** variety. The stepped cone has a constant angular velocity (or revolutions per unit time) that generates a different pulling speed at each capstan diameter. Figures 3.2 and 3.3 show, respectively, a schematic representation of a stepped cone drawing system,[17] and a stepped cone in a commercial drawing system.

It is fundamental that the drawing speed increases as the wire lengthens and is reduced in diameter in the upstream die. This is easily considered, since one can assume that the overall volume of the wire (equal to length multiplied by cross-sectional area) remains constant during its drawing. On this basis, the product of the drawing speed and the wire cross-sectional area remains constant.

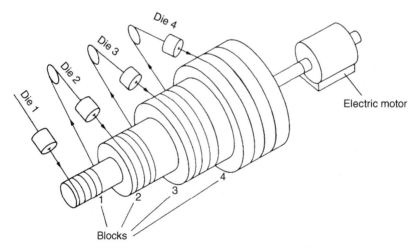

Figure 3.2 Schematic representation of a stepped cone drawing system. From J. N. Harris, *Mechanical Working of Metals*, Pergamon Press, New York, 1983, p. 208. Copyright held by Elsevier Limited, Oxford, UK.

Figure 3.3 A stepped cone within a commercial drawing system. (Courtesy of Macchine+Engineering S.r.l.)

In the case of the multiple-diameter capstan, the stepped diameters provide the series of drawing speeds consistent with the increased speed needed as the wire is reduced in cross section during the multiple-die drawing; that is, the wire is pulled through the first die by the smallest diameter on the capstan, goes through the second die, is pulled by the second smallest diameter on the capstan, and so on. The respective drawing speeds, v_1, v_2, v_3..., may be calculated from Equation 3.1 for the respective stepped cone diameters, D_1, D_2, D_3... with the value of ω remaining constant.

With separate capstans, the series of drawing speeds is achievable largely by driving the individual capstans at progressively higher angular velocities (values of ω). All capstans may be driven by a single power source, or the capstans may be driven individually. The angular velocity may be programmed and controlled so that the capstan surface speed is essentially the same as the intended drawing speed so that the wire does not slip on the capstan (**no slip machines**). Alternatively, the capstans may be driven faster so that the wire slips on the capstan by design (**slip machines**). Beyond this, wire speed may be controlled by variable "storage" of wire between passes on dynamic accumulating systems.

The design and operation of multiple-die machines vary significantly from ferrous to non-ferrous practice. With the ferrous drawing, conventional multiple-die drawing speeds reach 600 m/min, and, with non-ferrous drawing, speeds up to 2000 m/min are common. However, modern drawing machines have featured speeds several times these levels. The major limitation to such drawing speeds lies not in the drawing process, but in the dynamic equipment necessary to payoff, handle, and take up the wire. The frequency of wire breakage is an increasing consideration at high speed, since production may be lost while restringing the machine.

Some modern drawing machine systems also involve the drawing of several or many wires at once in parallel operation. With high drawing speeds and dozens of parallel lines, the productivity of these machines can be enormous. The basic principles of the individual drawing operations remain much the same, however. Important issues with such machines include string-up time, the amount of production lost due to wire breakage, the frequency of such breakage, and the cost and maintenance of the ancillary wire-handling equipment.

3.5. OTHER IN-LINE PROCESSES

Drawing is often done directly in line with other operations. These may include shaving (circumferential machining of the outer rod surface), descaling, pickling (chemical removal of surface oxide), cleaning, and the application of coatings and lubricants prior to initial drawing. Annealing and other thermal processes may be undertaken in tandem with drawing. Other in-line processes include numerous types of electrical insulation application, straightening, cutting, and welding. Finally, some drawing systems lead continuously to wire-forming operations (for fasteners, springs, etc.).

3.6. QUESTIONS AND PROBLEMS

3.6.1 A multiple-die wire drawing operation finishes at a diameter of 0.1 mm at a speed of 2,000 m/min. An upstream die has a size of 0.18 mm. What is the speed of the wire coming out of that upstream die? **Answer:** As stated in Section 3.4, the product of the drawing speed and the wire cross-sectional area remains constant. Therefore, final drawing speed multiplied by final area equals upstream speed multiplied by upstream area, and the upstream speed in question equals the final speed multiplied by the ratio of the final area to the upstream area. The area ratio can be replaced with the square of the diameter ratio. Therefore, Upstream speed $= (2000 \text{ m/min}) \times [(0.1)/(0.18)]^2 = 617 \text{ m/min}$.

3.6.2 If a stepped cone drawing machine lengthens the wire 26% in each drawing pass with five passes involved with no slip, and if the largest capstan diameter is 15 cm, what will be the smallest capstan diameter?

Answer: The percentage increase in length in each pass is associated with an identical increase in velocity, and an identical increase in associated capstan diameter (note Equation 3.1). After five passes, the velocity and associated capstan diameter will have increased by a factor of $(1.26)^5$, or 3.18. Thus, the final capstan diameter *divided* by 3.18 will be the diameter of the smallest, and first capstan. The diameter is 4.72 cm.

3.6.3 If the finishing speed in the previous problem is 1000 m/min, how many revolutions per minute is the capstan making?

Answer: Rearranging Equation 3.1, the revolutions per unit time, or ω, is $[v/(\pi D)]$. Thus, the number of revolutions per unit time is $(1000\,\text{m/min})/[(\pi)(0.15\,\text{m})]$, 2122 min^{-1}, or 35.4 s^{-1}.

3.6.4 A certain high-capacity, multi-line drawing machine is losing 4% of its productivity due to one drawing break each week. Assuming a 20-shift per week basis, with 7 active manufacturing hours per shift, estimate the time it takes to string up the machine.

Answer: The number of active manufacturing hours per week is 7 × 20, or 140 hours. Four percent of this number is 5.6 hours, or the time required to string up the drawing machine.

CHAPTER FOUR

Basic Engineering Variables Pertinent to Drawing

Contents

4.1. GENERAL QUANTITIES

4.1.1 Dimensions

The most encountered dimension in drawing is the **diameter, d**, of round bar, rod, or wire, and the corresponding die size. Such diameters are regularly given in **millimeters (mm)** or **inches (in.)**. One inch equals 25.4 mm, and it is important to be facile in both SI (International System) and UK/US measurement systems. Beyond this, wire diameters are frequently quoted in gage numbers, and a useful listing of common gage systems appears in the *CRC Handbook of Chemistry and Physics,* as well as in many

Wire Technology
ISBN 978-0-12-382092-1, DOI: 10.1016/B978-0-12-382092-1.00004-X

industrial handbooks.[18] Generally, higher gage numbers are associated with lower diameter values. For example, the American Wire Gage (AWG) is widely used with non-ferrous wire. In the AWG system, 12 gage implies a diameter of 0.0808 in. or 2.052 mm, whereas 24 gage implies a diameter of 0.0201 in. or 0.511 mm.

4.1.2 Force

The concept of **force**, particularly pulling force, is commonly used in drawing engineering. Fundamental definitions of force, such as the product of mass and acceleration, can be abstract and subtle as far as drawing is concerned, and are outside the scope of this text. Suffice it to say that drawing analyses generally involve steady-state dynamics where the drawing force is stable for a given pass and easily related to drawing speed, work, and power, as defined in the next section.

The SI unit for force is the **newton (N)** and its UK/US system counterpart is the **pound (lb)**, with one pound equal to 4.4482 N. Occasionally force is expressed in **kilograms (kg)**. This should be avoided, however, since the kilogram is not a unit of force, but of mass. When force data are expressed in kilograms (implying the force of gravity on a kilogram), they can be converted to newtons by multiplying by 9.8066 and converted to pounds by multiplying by 2.2046.

4.1.3 Work and Energy

Work is done, or **energy** expended, when a force is exerted through a distance. The SI unit for work and energy commonly used in drawing is the **joule (J)**, which is equivalent to a force of one newton exerted through a distance of one **meter (m)**. Work in the UK/US system is often expressed as **foot-pounds (ft-lb)**, with one foot-pound equal to 1.3558 J. Energy in the UK/US system is often expressed as **British thermal units (Btu)**, with one Btu equal to 1.0543×10^3 J.

4.1.4 Power

Power is the rate of work done, or energy expended, per unit time. The SI unit for power commonly used in drawing is the **watt (W)**, and a watt is equivalent to a joule expended in a second, or to a newton exerted through a meter in a second. This means that power can be viewed as force times speed, or drawing force times drawing speed, for a given pass.

Power in the UK/US system is often measured in **horsepower (hp)**, with one horsepower equivalent to 550 ft-lb/s or about 746 W. Power is often related to energy or work by multiplying the power, or average power, by the time the power is applied. Hence a watt-second would be a joule, and so on.

4.1.5 Stress

Stress is simply force divided by the area to which the force is applied. The SI unit for stress is the pascal (Pa), which is equivalent to a newton applied to a square meter of surface area. Since most stresses of interest in drawing are much larger, it is common to use megapascals (MPa), which are equal to 10^6 Pa. In the UK/US system, stress is usually expressed in **pounds per square inch (psi)** or 10^3 **pounds per square inch (ksi)**, where 1 ksi equals 6.894 MPa.

Drawing or pulling stresses are **tensile** stresses and are designated with a **positive** sign. Pushing stresses are called **compressive** stresses and are designated with a **negative** sign. However, pushing stresses are often called **pressures**, such as the die pressure in drawing. The signs of pressure are **positive**. Tensile and compressive stresses (or pressures) are called **normal** stresses, since the force is perpendicular to the surface. When the force is parallel to the surface, the stress is called a **shear** stress. Friction in drawing is an example of shear stress. Normal stresses will be designated with the Greek letter σ in this text, and shear stresses with the Greek letter τ.

Occasionally it is necessary to consider a change in the area to which the force is applied. A stress based on a "current" or "instantaneous" area is called a **true stress (σ_t)**, whereas a stress based on an initial area, disregarding any changes, is called an **engineering stress (σ_e)**.

4.1.6 Strain

When a workpiece is deformed, it is useful to relate the change in dimension, or the new dimension, to the original dimension. A **strain** is the ratio of the change in dimension to the original dimension. Strains are dimensionless, since we are dividing length by length, and so on. **Normal strains** (ε) involve changes in dimension that are parallel to the original dimension, such as occur in tension or compression. **Shear strains (γ)** involve changes in dimension that are perpendicular to the original, or reference dimension.

Tensile strains are widely used in drawing analysis. A tensile strain based on an original dimension (called **engineering strain)** will be designated by ε_e, where:

$$\varepsilon_e = (\ell_1 - \ell_0)/\ell_0, \qquad\qquad (4.1)$$

where ℓ_1 is the new length and l_0 is the original length. In drawing analyses, the strains to be considered are often large, and a measure of strain that recognizes the progressive changes in reference dimension is preferable. This strain measure is called **true strain**, and will be designated as ε_t, where:

$$\varepsilon_t = \ln(\ell_1/\ell_0), \qquad\qquad (4.2)$$

or the natural logarithm of the length ratio.

In drawing it is useful to consider that the volume of the workpiece remains constant, and that the product of the workpiece length and cross-sectional area remains constant, even though length increases and cross-sectional area decreases. Therefore:

$$A_0\,\ell_0 = A_1\ell_1 \text{ and } \ell_1/\ell_0 = A_0/A_1 \text{ and } \varepsilon_t = \ln(A_0/A_1), \qquad (4.3)$$

where A_0 is the original cross-sectional area and A_1 is the new cross-sectional area. Of course, the cross-sectional area of a round wire is just $(\pi/4)d^2$ and (A_0/A_1) is $(d_0/d_1)^2$, where d_0 is the original wire diameter and d_1 is the new wire diameter.

It is common practice to calculate drawing strains, as in Equations 4.1–4.3, in terms of the measured area or diameter change. Such strains assume **uniform** flow of the workpiece in drawing. In most cases, however, there is additional, non-uniform strain involved with passage through the die. Such strain is called **redundant**, and is introduced by multiplying the uniform strain by a **redundant strain factor**.

4.1.7 Strain rate

In drawing, the rate at which strain occurs in the workpiece can be important. The units of **stain rate** are 1/s or s^{-1}, and the average strain rate is simply given by the product of the drawing strain and drawing speed divided by the **length of the deformation zone, L_d** (see Section 4.2.4).

4.1.8 Relations between stress and strain

Below certain stress levels, stress and strain are related **elastically**, with stress proportional to strain and the elastic strain returning to zero when the stress is removed. The simplest relation of this kind is **Hooke's Law** for simple tension or compression:

$$\sigma = E\varepsilon, \qquad\qquad (4.4)$$

where **E** is **Young's modulus**. The units of Young's modulus are the same as those of stress.

There is a stress level, however, above which strain does not return to zero when stress is removed. Such remaining strain is called **plastic**, and the stress level is called the **yield strength (σ_y)**. Nearly all of the strain of interest in drawing is of the plastic type and the stress in the drawing zone is, in effect, at or above the yield strength of the workpiece. The strength that the bar, rod, and wire present during drawing will be called the flow stress designated σ_o.

4.1.9 Temperature

The **temperature** or temperature variation in the drawing process is of the utmost importance. An SI unit of temperature is **degrees Celsius (°C)**, and the basic UK/US unit of temperature is **degrees Fahrenheit (°F)**. As in the case of dimensions, it is important to be facile in both systems. One can convert from Fahrenheit to Celsius as follows:

$$°C = (°F - 32) \times (5/9) \qquad (4.5)$$

Some thermal analyses are based on absolute zero, or the lowest possible temperature (at which point thermal energy ceases). In the Celsius system this temperature is -273.15, and the **absolute temperature** in **degrees Kelvin (K)** is $°C + 273.15$. In the Fahrenheit system this temperature is -459.67, and the absolute temperature in **degrees Rankine (°R)** is $°F + 459.67$.

4.2. QUANTITIES DESCRIBING THE WORKPIECE AND DIE DURING DRAWING

4.2.1 Overview

Figure 4.1 is an analytical schematic of the wire drawing operation, with drawing proceeding left to right. The meanings of the several parameters are set forth in the next section.

4.2.2 Cross-sectional areas and the reduction

The wire enters the die at the left in Figure 4.1 with a **cross-sectional area** designated as A_0, and exits to the right with a cross-sectional area designated as A_1. The **reduction, r**, is

$$r = (A_0 - A_1)/A_0 = 1 - (A_1/A_0) \qquad (4.6)$$

or, in percent:

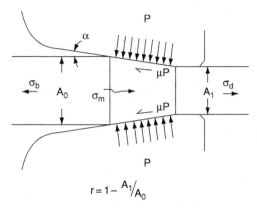

Figure 4.1 An analytical schematic illustration of a drawing pass.

$$r = [(A_0 - A_1)/A_0] \times 100 = [1 - (A_1/A_0)] \times 100 \qquad (4.7)$$

4.2.3 Die angle

In the case of round wire, the die (in the great majority of die designs) imposes converging flow upon the wire by way of a straight cone with a **die angle**, α. This region of converging flow is called the **drawing channel**. Note that α is the angle between the die wall and the drawing centerline, sometimes actually called the **"half-angle"** or **"semi-angle,"** with the **"included angle"** of the die as 2α. Die half-angles and included angles are usually given in **degrees**. However, in certain analytical expressions, the angles are expressed in **radians (rads)**, with 1 rad equal to $360/(2\pi)$, or 57.30, degrees.

When the drawing of shaped cross sections is undertaken, differing die angles will present themselves, depending on the orientation of the long-itudinal section.

4.2.4 Deformation zone shape and Δ

Figure 4.1 reveals a trapezoidal zone bounded by (a) lines perpendicular to the drawing axis, where the wire first makes contact and last makes contact with the die wall; and (b) the lines of the die wall. This is the nominal plastic **deformation zone**, and the shape of this zone, dependent on r and α, is fundamental to drawing analysis. The shape of the deformation zone is characterized by the ratio Δ, where:

$$\Delta = (\text{average height of zone}, \perp \text{to drawing axis})/$$
$$(\text{length of zone}, //\text{to drawing axis}). \qquad (4.8)$$

An approximate numerical value for Δ can be calculated from the relationships:

$$\Delta \approx (\alpha/r)\left[1 + (1-r)^{1/2}\right]^2 \approx 4 \tan \alpha / \ln[1/(1-r)] \qquad (4.9)$$

Table 4.1 displays Δ values for die angles and reductions of common interest in drawing. In general, low Δ values are associated with low die angles and high reductions, and high Δ values are associated with high die angles and low reductions. Figure 4.2 displays three different deformation zones and the associated values of Δ, α, and r.

The **length of the deformation zone (L_d)**, parallel to the drawing axis is $(d_0 - d_1)/(2 \tan\alpha)$ and the length of contact along the die wall, or **die contact length (L_c)**, is $(d_0 - d_1)/(2 \sin\alpha)$.

Sophisticated analyses of drawing make it clear that the shape of the actual deformation zone, in longitudinal section, is more complicated than a

Table 4.1 Values of Δ as a function of percent reduction and die semi-angle

Die Semi-Angle(°)	5%	10%	15%	20%	25%
4	5.5	2.7	1.7	1.3	1.0
6	8.2	4.0	2.6	1.9	1.5
8	10.9	5.3	3.5	2.5	2.0
10	13.7	6.7	4.3	3.1	2.4

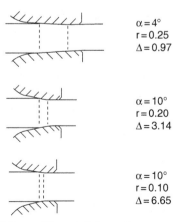

$\alpha = 4°$
$r = 0.25$
$\Delta = 0.97$

$\alpha = 10°$
$r = 0.20$
$\Delta = 3.14$

$\alpha = 10°$
$r = 0.10$
$\Delta = 6.65$

Figure 4.2 Illustrations of three different deformation zones and associated values of Δ, die semi-angle, and reduction.

simple trapezoid. However, the trapezoidal deformation zone geometry is the basis of process design and practical analysis.

4.2.5 Drawing stress and back stress

The symbol σ_d at the right of Figure 4.1 represents the **drawing stress** or the drawing or pulling force divided by A_1. In many cases there is a **back stress** in the opposite direction, where the wire enters the die, which is represented by σ_b at the left in Figure 4.1 and equal to the back force divided by A_0.

4.2.6 Die stresses

Figure 4.1 indicates the average **die pressure**, **P**, acting upon the wire in the deformation zone. This pressure represents the total normal force acting between the wire and the die divided by the area of contact between the wire and the die. While most analyses incorporate the average die pressure as though it were constant or uniform, sophisticated drawing analyses indicate that the pressure is not uniform, but higher near the drawing channel entrance and exit and lower in between.

Figure 4.1 indicates an average **frictional stress**, **μP**, where μ is the average **coefficient of friction**. The frictional stress is in the opposite direction of wire motion at the die wall, and is equal to the frictional force divided by the area of contact between the wire and the die. It is understood that frictional stresses will vary within the drawing channel, but this is rarely taken into consideration in practical analysis.

4.2.7 Centerline stress

Figure 4.1 indicates a value of stress, σ_m, near the center of the deformation zone. This is the average or **mean normal stress at the centerline**, particularly at the point where σ_m has the most tensile, or least compressive, value. Tensile values of σ_m are of great concern in drawing because they can lead to fracture at the wire center.

4.3. QUESTIONS AND PROBLEMS

4.3.1 Wire is drawn through a certain die with a drawing force of 500 N, at a speed of 1000 m/min. How much power is consumed in watts and in horsepower?

Answer: Power is equal to force multiplied by speed, or $(500 \text{ N}) \times (1000 \text{ m/min}) \times (1 \text{ min}/60 \text{ s})$, which equals 8333 Nm/s, 8333 W, or 11.2 hp.

4.3.2 Consider the strain in a 1 AWG reduction based on diameter change. Calculate its value as true strain and as engineering strain.

Answer: As noted in Section 1.2.2, a 1 AWG pass involves an area reduction of about 20.7%. Therefore the ratio (A_0/A_1) is $1/(0.793)$ or 1.261, and this is also the ratio (ℓ_1/ℓ_0). From Equation 4.2, true strain is the natural logarithm of (ℓ_1/ℓ_0) or 0.232. Equation 4.1 can be rewritten so that engineering strain is $[(\ell_1/\ell_0) - 1]$ or 0.261.

4.3.3 A rod is drawn from a diameter of 6 to 5.5 mm with a die semi-angle of 6 degrees at a speed of 200 m/min. What is the strain rate based on diameter change?

Answer: Strain rate is the product of strain and speed divided by the length of the deformation zone. The true strain in this case is the natural logarithm of $[6/(5.5)]^2$ or 0.174. From Section 4.2.4, the deformation zone length can be calculated as $(d_0 - d_1)/(2 \tan\alpha)$, or 2.38 mm. Therefore, the strain rate is $(0.174) \times (200 \text{ m/min}) \times (1 \text{ min}/60 \text{ s}) \div (0.00238 \text{ m})$, or 244 s^{-1}.

4.3.4 A 20% reduction is taken with a die semi-angle of 6 degrees. A 15% reduction with no change in Δ is taken. What should the new die semi-angle be?

Answer: Using Equation 4.9, Δ can be calculated as 1.88 (compare to 1.9 in Table 4.1). Since Δ does not change in this case, Equation 4.9 can be rearranged as $\alpha \approx \Delta \, r \, [1 + (1 - r)^{\frac{1}{2}}]^{-2}$. On this basis α is 0.0765 or 4.38°.

CHAPTER FIVE

Basic Drawing Mechanics

Contents

5.1. A SIMPLE DRAWING STRESS MODEL

5.1.1 Drawing stress and work per unit volume

As noted in Section 4.1.3, work is simply the product of force times distance. So, if a drawing force, F, is applied to pull a length of wire or rod, L, downstream from the die, the work done, W, is simply:

$$W = F \times L. \tag{5.1}$$

Now, if L is shifted to the left side of this equation, and both sides are divided by A_1, the cross-sectional area of the wire downstream from the die, then

$$W/(L \times A_1) = F/A_1. \tag{5.2}$$

The term on the left is simply work divided by volume or **work per unit volume**, **w**, and the term on the right is simply the draw stress, σ_d. Thus we have the major conclusion that the drawing stress equals the work expended per unit volume, or

Wire Technology
ISBN 978-0-12-382092-1, DOI: 10.1016/B978-0-12-382092-1.00005-1

$$\sigma_d = w. \tag{5.3}$$

This analysis can be considered *external*, since the terms involved are defined and measured outside the drawing die and deformation zone. The *same* work expended in drawing can be accounted for, however, by an *internal* analysis, which addresses **uniform work, w_u, non-uniform** or **redundant work, w_r,** and **friction work, w_f,** all on a per-unit-volume basis. Each of these internal work-per-unit-volume terms contributes directly to the drawing stress, and a drawing stress formula that reflects these terms is very useful in analyzing drawing behavior. Thus, we will model the drawing stress on the basis of

$$\sigma_d = w_u + w_r + w_f. \tag{5.4}$$

5.1.2 Uniform work

Uniform work is the work done in stretching (or thinning) the wire or rod without consideration of the interaction with the die; that is, the effects of the die in changing the direction of metal flow (convergence) or in exerting frictional forces are not considered.

Uniform work per unit volume is equivalent to the work per unit volume that would be expended in a tensile test, up to the strain of the drawing reduction modeled, provided necking or fracture did not intervene. That work-per-unit-volume value would be obtained by integrating the stress times the strain differential, or determining the area under the true stress–true strain curve. This is most simply expressed as the product of the **average flow stress in the wire**, during drawing, σ_a, and the true strain of the drawing reduction; that is,

$$w_u = \sigma_a \ln (A_0/A_1) = \sigma_a \ln [1/(1-r)]. \tag{5.5}$$

While the *average* flow stress is employed in simpler analyses, it should be understood that the flow stress can be expected to increase from die entry to die exit due to strain hardening.

5.1.3 Non-uniform or redundant work

When the incoming wire contacts the die wall, it is deflected from the "horizontal" toward the centerline, consistent with the die half-angle, α. In an opposite manner, the wire, upon leaving the drawing cone, is deflected back to the horizontal from its previous trajectory parallel to the die wall. The changes in velocity involved are shown in Figure 5.1. In the simplest analysis, the wire can shear at the entrance to the deformation zone, and

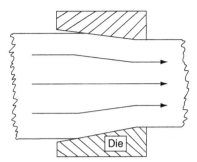

Figure 5.1 Flow of the workpiece through the drawing die. Note the deflections upon entering and exiting the deformation zone.

shear in an opposite direction at the exit to the deformation zone. These opposite shears cancel one another, and are thus not reflected in the overall drawing reduction of the wire. These strains are often called redundant strains, because there is no net effect demonstrable in the overall wire geometry. The related work does not cancel, and this work, divided by volume, is called redundant work.

It has proven difficult to model redundant work from first principles, and semi-empirical approaches from generalized physical measurements have received the most application. The most common, a **redundant work factor**, Φ, is employed, where

$$\Phi = (w_u + w_r)/w_u, \tag{5.6}$$

or where

$$w_r = (\Phi - 1)w_u = (\Phi - 1)\sigma_a \ln[1/(1 - r)]. \tag{5.7}$$

Obviously, if there is no redundant work, then $\Phi = 1$. In general, redundant work does exist, especially at Δ values above one or with higher die angles and/or lower drawing reductions. The most useful classical data are those of Wistreich,[19,20] which, for typical drawing passes, lead to the relation

$$\Phi \approx 0.8 + \Delta/(4.4). \tag{5.8}$$

Note that redundant work increases substantially as Δ increases.

The nature of redundant work and its relationship to Δ may be seen in Figure 5.2a and b.[21] Figure 5.2a displays hardness profiles for the upper half of a longitudinal section of a copper rod. The hardness is expressed in terms of the Knoop micro-hardness scale (KHN). Hardness (see Section 11.5) is resistance to penetration and correlates well with strength. The left portion of Figure 5.2a displays lines of constant hardness in the deformation zone.

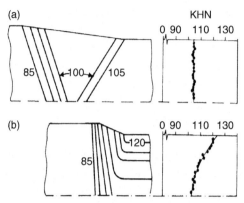

Figure 5.2 Longitudinal section iso-hardness lines and hardness profiles for (a) a low angle (low Δ) drawing pass and (b) a high angle (high Δ) drawing pass. The radial hardness gradient in (b) reflects the presence of redundant work. From W. A. Backofen, *Deformation Processing*, Addison-Wesley Publishing Company, Reading, MA, 1972, p. 139. Copyright held by Pearson Education, Upper Saddle River, NJ, USA.

Note that the rod enters the deformation zone at KHN 85 and exits at KHN 105. The right portion of Figure 5.2a reveals that the rod has been uniformly hardened to KHN 105 from centerline to surface. The uniform hardening from KHN 85 to 105 reflects the strain hardening experienced by the copper.

The value of Δ implied in Figure 5.2a is about 1.3, and in a range where strain and related strain hardening are rather uniform. In Figure 5.2b, the value of Δ appears to be about 5.3, and the strain is non–uniform and the redundant strain is extensive in the outer radius region. The centerline region displays no hardening beyond the KHN 105 level associated with nearly uniform strain in Figure 5.2a. The surface hardness reaches a value of at least KHN 120. This increased surface hardness and related non–uniform strain reflect redundant work. The Δ value of Figure 5.2b implies, from Equation 5.8, that the redundant work factor Φ is about two, and that the value of w_r is about the same as w_u. In other words, the deformation work implied in Figure 5.2b is about twice that implied in Figure 5.2a. The large amount of extra strain and deformation work, concentrated as it is nearer the rod surface, significantly strengthens the rod in a way that is especially noticeable in bending and torsion and threading and annealing operations.

5.1.4 Friction work

The contribution of friction to the drawing stress, or the frictional work divided by the wire or rod volume, may be obtained by multiplying the contact area between the wire and the drawing cone by the frictional shear stress acting on that area and then dividing by A_1. As noted in Section 4.2.6, the average frictional stress is commonly expressed as μP, where μ is the average coefficient of friction and P is the average die pressure. The average die pressure can be approximated as $\Phi\sigma_a$, with die pressure reflecting redundant as well as uniform work in the deformation zone. The overall expression for frictional work per unit volume is

$$w_f = \mu \cot\alpha \, \Phi\sigma_a \ln[1/(1-r)]. \qquad (5.9)$$

Now, since Δ may be approximated by $4\tan\alpha/\ln[1/(1-r)]$, Equation 5.9 may be rewritten as

$$w_f = 4\mu \, \Phi\sigma_a/\Delta. \qquad (5.10)$$

It might be argued that additional frictional work occurs in the cylindrical section immediately downstream from the drawing cone (called the **bearing** or **land**, as per Section 9.3.2), as shown in Figure 4.1. However, there is no theoretical basis for contact in this area, apart from that associated with misalignment of the wire and die axes. Accordingly, such work is generally omitted in practical analysis.

5.1.5 The drawing stress formula[20]

Placing the terms of Equations 5.5, 5.7, and 5.10 into Equation 5.4, one obtains

$$\sigma_d = w_u + w_r + w_f = \sigma_a \ln[1/(1-r)] + (\Phi-1)\sigma_a \ln[1/(1-r)] + 4\mu \, \Phi\sigma_a/\Delta. \qquad (5.11)$$

Combining terms, dividing through by σ_a, utilizing Equation 4.9, and using $\tan\alpha \approx \alpha$, one obtains

$$\sigma_d/\sigma_a = (4\Phi/\Delta)(\alpha+\mu). \qquad (5.12)$$

Finally, using Equation 5.8 and using Σ for the drawing stress ratio, σ_d/σ_a, one obtains

$$\sigma_d/\sigma_a = \Sigma = [(3.2/\Delta)+0.9](\alpha+\mu). \qquad (5.13)$$

5.2. DRAWING LIMITS

The drawing stress must remain below the flow stress at the die exit, σ_{01}, to avoid uncontrolled stretching, necking, or fracture. Where σ_{00} is the flow stress at the die entrance, the average flow stress can be approximated as

$$\sigma_a = (\sigma_{00} + \sigma_{01})/2. \tag{5.14}$$

Thus, for the case of no strain hardening, $\sigma_a = \sigma_{00} = \sigma_{01}$, σ_d must remain below σ_a, and Σ must remain below one. Of course, in the case of significant strain hardening, σ_d can theoretically exceed σ_a, since the increased strength at the die exit allows for a higher drawing stress without uncontrolled stretching, necking, or fracture. However, for general and conservative analysis, it will be assumed that σ_d cannot exceed σ_a, and a drawing limit of $\Sigma = 1$ will be recognized.

In a theoretical case where there is no redundant work and no friction work, Equations 5.4 and 5.5 indicate that the drawing limit occurs when ln $[1/(1 - r)] = 1$, or when r = 0.632 or 63.2%. Of course, practical drawing passes involve redundant work and friction work, and drawing limits involve reductions well below 63.2%. As noted in Section 1.2.2, commercial practice rarely involves reductions above 30 or 35%, and much lower reductions are common.

5.3. AN ILLUSTRATIVE CALCULATION

Let us consider a drawing pass involving a single AWG reduction of 20.7% taken through a die with a semi-angle of 8° or 0.14 rad (the values of α in Equation 5.13 must be in radians). Equation 4.9 thus indicates a Δ value of 2.42.

Let us further consider that the average coefficient of friction is 0.1, as indicative of "bright" drawing practice with a liquid lubricant (see Section 8.1.3).

Placing these values into Equation 5.13 results in a Σ value of 0.53, well below the drawing limit of 1.0. Since a pass such as this is quite common in drawing practice, it is clear that many drawings are undertaken at drawing stress levels well below those that would normally be associated with uncontrolled stretching, necking, or fracture. Thus, when such failures are encountered, extraordinary conditions are in play.

5.4. THE ISSUE OF OPTIMUM DIE ANGLES AND Δ VALUES

Equation 5.8 indicates that redundant work and its contribution to the drawing stress *increase* as Δ increases. Equation 5.10 indicates that friction work and its contribution to the drawing stress *decrease* as Δ increases. These dependencies lead to a minimization of drawing stress at an intermediate or optimum value of Δ. If the right side of Equation 5.13 is converted to an expression involving only Δ, μ, and r, one obtains the relation

$$\sigma_d/\sigma_a = \Sigma = [(3.2/\Delta) + 0.9]\{\Delta r[1 + (1-r)^{1/2}]^{-2} + \mu\}. \qquad (5.15)$$

Differentiating the right-hand side of Equation 5.15 with respect to Δ, and setting the derivative equal to zero, the value of Δ associated with the minimum value of σ_d/σ_a or Σ can be solved. In a similar manner, the value of α associated with the minimum value of σ_d/σ_a or Σ can be determined. These optimum values are

$$\Delta_{opt} = (1.89)(\mu/r)^{1/2}[1 + (1-r)^{1/2}], \qquad (5.16)$$

$$\alpha_{opt} = (1.89)(\mu r)^{1/2}/[1 + (1-r)^{1/2}]. \qquad (5.17)$$

Figure 5.3 shows an experimental determination of the minimization of σ_d/σ_a or Σ as a function of die angle for a wide range of reductions.[19]

It must be emphasized, however, that there are many other factors to consider when selecting a value of die angle or Δ for a drawing pass. Moreover, the value of the drawing stress varies only slightly over a wide range of die angle or Δ values. Beyond this, low Δ drawing passes often facilitate lubrication and lower the value of μ, thus pushing the value of Δ_{opt} or α_{opt} to lower values.

Thus, while the concept of an optimum die angle or Δ value has important tutorial value in understanding friction and redundant work, it is rarely the sole basis for selecting a value of die angle or Δ for a drawing pass.

5.5. DIE PRESSURE

The average die pressure, P, (see Section 4.2.6) reflects, for the most part, the uniform and redundant work. It is often useful, then, to consider the ratio, P/σ_a, of average die pressure to average flow stress, and to express this ratio as a function of Φ or Δ. Figure 5.4 displays data from Wistreich that show relationships between Δ and P/σ_a.[19,20] When Δ is near one, the average die pressure is within $\pm 20\%$ of σ_a. For higher values of Δ, the average die pressure increases linearly with Δ, and may be approximated as

$$P/\sigma_a = \Delta/4 + 0.6. \qquad (5.18)$$

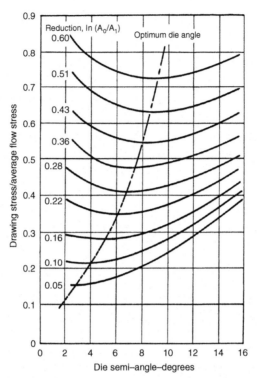

Figure 5.3 Experimental determination of the dependency of (σ_d/σ_a) as a function of die angle, for a wide range of reductions. From J. G. Wistreich, Proceedings of the Institution of Mechanical Engineers, 169 (1955), p. 659. Copyright held by Professional Engineering Publishing, London, UK.

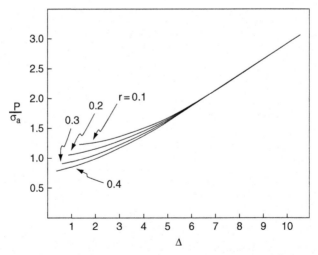

Figure 5.4 Ratio of average die pressure to average flow stress as a function of Δ. Data from J. G. Wistreich, *Proceedings of the Institution of Mechanical Engineers*, 169 (1955), p. 654 and R. N. Wright, Wire Technology, 4(5), 1976, p. 57.

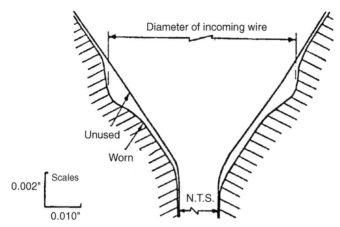

Figure 5.5 Profile of a worn die. Note difference in vertical and horizontal scales. From J. G. Wistreich, *Metallurgical Reviews*, 3 (1958), 97. Copyright held by Maney Publishing, London, UK, www.maney.co.uk/journals/mr and www.ingentaconnect.com/content/maney/mtlr.

In the lower Δ range, lighter reductions result in increased values of die pressure over and above the role of smaller reductions in increasing Δ.

The die pressure does not vary greatly with friction in the practical range. High levels of friction will, however, substantially *decrease* die pressure.

Increased die wear is often associated with increased die pressure and increased Δ. This is the case at the locus where the wire or rod first contacts the die. At this point, vibrations and fluctuations in alignment, diameter, and lubrication cause variations and cycling in the initial application of pressure to the die. This results, in effect, in fatigue failure of the die material and the development of a "wear ring" at the initial contact locus. Figure 5.5 shows the profile of such a wear ring.[9]

5.6. CENTERLINE TENSION

The deformation zone is bounded by a tensile stress (drawing stress) along the die exit and by frictional shear stress and compressive stress (die pressure) along the die wall. At the entry, there is either no stress or a tensile back stress.

The stress state is far from uniform in the deformation zone, however. The average stress at the centerline is less compressive than at the die wall and may even be tensile. This is particularly the case for higher values of Δ. Figure 5.6 displays values, calculated by Coffin and Rogers[22,23] from the work of Hill and Tupper,[24] of the ratio of the average or hydrostatic

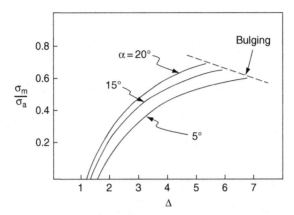

Figure 5.6 The ratio of centerline hydrostatic stress to average flow stress, as a function of Δ. From L. F. Coffin, Jr., and H. C. Rogers, *ASM Transactions Quarterly*, 60 (1957) 672; H. C. Rogers and L. F. Coffin, Jr., *International Journal of Mechanical Science*, 13 (1971) 141; and R. Hill and S. J. Tupper, *Journal of the Iron and Steel Institute*, 159 (4) (1948) 353.

centerline stress, σ_m, to the average flow stress, σ_a, as a function of Δ. Below a Δ value of about 1.3, the centerline hydrostatic stress is compressive. Above a Δ value of about 1.3, σ_m becomes increasingly tensile with increased Δ, reaching values well above half the average flow stress. In general, higher die angles result in increased values of centerline tension over and above the role of higher die angles in increasing Δ.

Centerline tension in drawing is of great concern because it promotes the development and growth of porosity and ductile fracture of the wire at the centerline. Moreover, it inhibits the stabilization and densification of porous centerline structure inherited from casting operations. Fractures developing at the centerline during drawing are called "center bursts," "central bursts," and "cuppy cores." This author has published a practical review of the development of center bursts.[25] Figure 5.7 displays a longitudinal section of gross center bursts cited by Kalpakjian from the work of Breyer.[26]

5.7. PLASTIC FLOW OUTSIDE THE DRAWING CONE

Most analyses of drawing mechanics assume that deformation is confined to the trapezoidal zone in the longitudinal section of Figure 4.1, and this is the shape that is the basis for Δ. However, for higher Δ values plastic deformation occurs both up- and downstream from this nominal

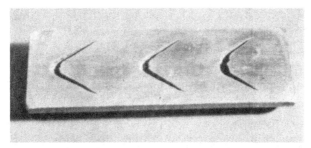

Figure 5.7 Longitudinal section displaying gross center bursting. From S. Kalpakjian, *Mechanical Processing of Materials*, D. Van Nostrand Company, New York, 1967, 175. Copyright held by S. Kalpakjian, Boise, ID, USA.

deformation zone, as shown in the measurements of Wistreich displayed in Figure 5.8.[19]

The upstream deformation is generally called "bulging," and can be thought of as a "rejection" of near-surface metal by higher die angles. If the die angle is high enough, metal will be torn or cut away from the wire or rod surface. Such cutting behavior is a class of "shaving," although practical shaving processes generally involve a semi-angle greater than $\pi/2$ (90°). In the absence of actual shaving, the near-surface metal does enter the die, involving gross redundant work and impedance of lubricant entry into the die. Figure 5.6 displays a threshold for bulging, as a function of Δ and die

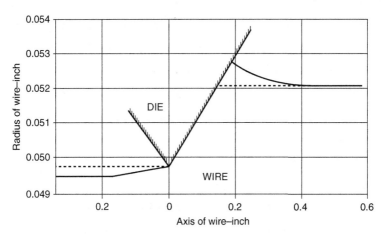

Figure 5.8 Longitudinal section profile displaying deformation occurring up- and downstream from the nominal deformation zone. Note difference in vertical and horizontal scales. From J. G. Wistreich, *Proceedings of the Institution of Mechanical Engineers*, 169 (1955) 659. Copyright held by Professional Engineering Publishing, London, UK.

semi-angle. The conditions for the behavior displayed in Figure 5.8 correspond to a Δ value of 14.25 and a die semi-angle of 0.27 (15.5°), well beyond the bulging threshold.

Downstream deformation has been referred to as "suck down." It must be understood that the viability of the drawing process depends on the drawing stress remaining clearly below the yield stress or flow stress of the exiting wire. Hence suck down is not to be viewed as simple pulling strain due to the draw stress. Instead, it is an extension of plastic flow beyond the nominal deformation zone shown in Figure 4.1. In the exemplary measurements of Figure 5.8, the plastic flow has occurred nearly two wire diameters beyond the drawing cone exit, and the extent of the wire area reduction is a non-trivial 1.2%. Of course, with more typical Δ values, this value will be much less and possibly within the uncertainties of die size and the effects of wear and misalignment. Any suck down at all implies that perfect wire/die alignment involves no pressure in the bearing or land cylinder immediately downstream from the drawing cone. In actual practice, such alignment is not always possible, and rubbing of the wire on the bearing can be expected on one side or the other.

 ## 5.8. EFFECTS OF BACK TENSION

The concept of back tension, σ_b, was introduced in Section 4.2.5, as per Figure 4.1, and this author has published a review of die and pass schedule design implications.[27] Clearly the back tension must add to the overall drawing stress, and it can be shown from simple work and force balance arguments that, *in the absence of friction,*

$$\sigma_d = \sigma_{do} + \sigma_b. \tag{5.19}$$

where σ_{do} is the draw stress that would be present if there were no back tension.

One of the effects of back tension is to lower the average die pressure, P, and this lowers the average frictional stress, μP. So, *in the presence of friction,* σ_d is lower than expressed by Equation 5.19. Analysis consistent with the approach of this chapter leads to the expression

$$\sigma_d = \sigma_a[(3.2/\Delta) + 0.9](\alpha + \mu) + \sigma_b[1-(\mu r/\alpha)(1 - r)^{-1}]. \tag{5.20}$$

The effect on the average die pressure, P, can be expressed as follows:

$$P/P_o = 1 - [2b/(2 - \Sigma)], \tag{5.21}$$

where P_o is the average die pressure in the absence of back tension and b is σ_b/σ_a. As introduced in Sections 5.1.2 and 5.1.5, σ_a is the average flow stress and Σ is σ_d/σ_a.

In addition to increasing overall drawing stress and lowering die pressure and related friction, back tension adds directly to the value of centerline tension increasing the risk of fracture at the wire center. Nonetheless it may, on occasion, be useful to add practical levels of back tension to reduce die pressure and related friction. In such cases, practical analysis shows that increased risk of centerline damage can be largely avoided if Δ is decreased by about 0.75.

5.9. QUESTIONS AND PROBLEMS

5.9.1 What are the Δ values for a 20% drawing pass with included angles of 20°, 16°, 12°, and 8° ?

Answer: Note that the included angle is twice the value of the die semi-angle. Use Table 4.1 to determine respective Δ values of 3.1, 2.5, 1.9, and 1.3 for the included angles of 20°, 16°, 12°, and 8°. Alternatively, Equation 4.9 can be used making sure to convert the die semi-angle to radians.

5.9.2 What are the values of the drawing stress, σ_d, for the four cases in problem 5.9.1, assuming a coefficient of friction of 0.10 and an average strength of 350 MPa? What would the values of the drawing force be for a 0.5 mm as-drawn wire diameter?

Answer: Using Equation 5.13 and recognizing that the die semi-angle must be in radians, the respective draw stresses are 186, 183, 185, and 200 MPa for the included die angles of 20°, 16°, 12°, and 8°. The as-drawn cross-sectional area is $0.196\,\text{mm}^2$, thus the respective drawing forces are 36.5, 35.9, 36.4, and 39.3 N for the included die angles of 20°, 16°, 12°, and 8°. Note the rather modest variation in drawing stress and force, despite the wide range of die angles.

5.9.3 What are the average die pressures for the four cases in Problem 5.9.1, assuming an average strength of 350 MPa?

Answer: Using Equation 5.18, the respective die pressures are 502, 429, 376, and 324 MPa for the included die angles of 20°, 16°, 12°, and 8°. Note the significant drop in die pressure as the die angles and Δ values decrease.

5.9.4 What are the redundant work factors for the four cases of Problem 5.9.1?

Answer: Using Equation 5.8, the respective redundant work factors are 1.50, 1.37, 1.23, and 1.10 for the included die angles of 20°, 16°, 12°, and

8°. Note the significant drop in redundant work as the die angles and Δ values decrease.

5.9.5 What are the values of centerline tension for the four cases in Problem 5.9.1, assuming an average strength of 350 MPa?

Answer: Consult Figure 5.6, noting that the die semi-angle has an effect above and beyond its role in Δ. The respective centerline tension values are about 140, 97, 41, and <0 MPa for the included die angles of 20°, 16°, 12°, and 8°. Note the significant drop in centerline tension (leading ultimately to compression) as the die angles and Δ values decrease.

5.9.6 What are the values of drawing stress, σ_d, for a 20% drawing pass with an included angle of 12° and coefficients of friction of 0.04, 0.10, 0.15, and 0.25? Assume an average strength of 350 MPa. Are all of these drawing conditions possible when comparing drawing stress with average strength?

Answer: Use Equation 5.13 and convert included die angle to die semi-angle. The respective drawing stresses are 131, 185, 231, and 321 MPa for the coefficients of friction of 0.04, 0.10, 0.15, and 0.25. Friction, of course, adds to the drawing stress; however, all four drawing conditions seem possible since σ_d remains below σ_a.

5.9.7 What is the largest reduction that can be taken, in one pass, with an included angle of 12° and a coefficient of friction of 0.10?

Answer: Use Equation 5.13 and set Σ equal to one. The value of Δ can be calculated to be 0.804. Using Equation 4.9, the value of the (largest) reduction can be calculated to be about 0.41 or 41%. However, the calculation for the reduction does require iteration.

5.9.8 A pass from 1.00 to 0.90 mm with a 12° included die angle displays a drawing force of 200 N for a wire with an average strength (under drawing conditions) of 700 MPa. Estimate the coefficient of friction.

Answer: The as-drawn cross-sectional area can be calculated as 0.636 mm^2 and the drawing stress is 314 MPa. The reduction may be calculated as 0.19 or 19%, and the value of Δ is 2.0 for the 6° (0.105 rad) die semi-angle. Using Equation 5.13, the coefficient of friction can be calculated to be 0.074.

CHAPTER SIX

Drawing Temperature

Contents

6.1. CONTRIBUTIONS TO THE DRAWING TEMPERATURE

6.1.1 The temperature of the incoming wire

The temperature of the wire just prior to die entry, or T_0, is perhaps the most important consideration in drawing temperature analysis, since in tandem drawing it may reflect a large amount of undissipated heat from previous drawing passes; that is, the wire is unlikely to cool back to ambient temperature in going from one pass to the next. Beyond this, it may be desirable to pre-heat or -cool the wire before drawing. Wire that has been stored at ambient shop or warehouse temperature will reflect that temperature (plus normally modest heating effects of uncoiling, straightening, or surface conditioning) upon die entry. In any event, all drawing temperature projections include a prominent role of T_0.

Wire Technology
ISBN 978-0-12-382092-1, DOI: 10.1016/B978-0-12-382092-1.00006-3

6.1.2 The overall temperature increase in the drawing pass

The work per unit volume done in drawing, w, is nearly all converted to heat. It may be argued that a small portion results in increased crystal defect energy in the wire (dislocations, high energy dislocation configurations, vacancies, and other forms of "stored energy"), and in externally dissipated vibrations and inertial responses. However, most practical drawing analysis ignores these possibilities, and relates the drawing work per unit volume directly to increases in wire and die temperature. Moreover, the work per unit volume, w, is equivalent to the drawing stress, σ_d.

In start-up drawing or in the drawing of short lengths of wire and rod, the "cold" die may absorb a significant fraction of the heat. However, in steady-state commercial drawing, internal die surface temperatures become quite high, and heat flow from the wire into the die has been shown to be minimal. Thus, simple analyses of heating in commercial drawing assume, as a good approximation, that all of the work per unit volume results in heating of the wire.

As the wire exits the die there will be a thermal gradient because of frictional heating; that is, the wire will be hotter at the surface than at the center. This gradient is short-lived, however, and the wire shortly equilibrates to a relatively uniform temperature, except for the effect of convective surface cooling. This equilibrated temperature, $\mathbf{T_{eq}}$, can be expressed as

$$T_{eq} \approx T_0 + \sigma_d/(C\rho), \qquad (6.1)$$

where C is the specific heat of the wire and ρ is the density of the wire.

The distance downstream from the die exit where equilibration should occur, $\mathbf{L_{eq}}$, can be estimated from basic heat transfer theory as

$$L_{eq} \approx (vC\rho d^2)/(24K), \qquad (6.2)$$

where $\mathbf{K}$ is the thermal conductivity of the wire. Table 6.1 provides values of C, ρ, and K for commonly drawn metals, as well as a highly useful conversion factor.

Table 6.1 Physical values for thermal calculations for four commonly drawn metals

	Aluminum	Copper	Carbon Steel	Tungsten
Specific heat in J/(g °C)	0.896	0.385	0.461	0.151
Density in g/cm^3	2.7	8.9	7.8	19
Thermal conductivity in J/(cm s °C)	2.01	3.85	0.461	1.17

Note: In making thermomechanical calculations, it is important to remember that an MPa equals J/cm^3.

6.1.3 Illustrative calculations

First, consider a hypothetical aluminum wire drawing pass with the values shown in Table 6.2. The value for the drawing stress, σ_d, can be obtained from Equation 5.13, first calculating Δ by way of Equation 4.9. In the case of this hypothetical aluminum drawing pass, Δ is 2.42 and σ_d is 85 MPa. Inserting σ_d into Equation 6.1, using the values of C and ρ from Table 6.1 and assuming a T_0 of 20°C, a value for T_{eq} of 55°C is obtained. That is, the wire enters the die at a wire temperature of 20°C and exits the die at an equilibrated temperature of 55°C with the work per unit volume of drawing increasing the temperature by 35°C.

The increase of 35°C in one pass is significant for a lower melting point metal such as aluminum. Moreover, this temperature increment may be added to the wire at each pass. Thus, considerable heat must be removed between passes to avoid a high and uncontrollable drawing temperature. This may not be easy with the oil-based lubricants generally used with aluminum (water functions much better as a coolant than oil).

Placing the values from Tables 6.1 and 6.2 into Equation 6.2, an L_{eq} value for the hypothetical aluminum pass of 0.25 cm can be calculated. This is equal to five times the wire diameter, and is a small value in relation to the distance to the subsequent capstan contact. Hence, T_{eq} is a reasonable estimate for general wire temperature prior to cooling from capstan contact or the convective flows of air and lubricant.

Table 6.2 also provides information for a hypothetical carbon steel drawing pass. Using this data and that of Table 6.1, Δ and σ_d values for the steel drawing pass of 1.08 and 330 MPa, respectively, can be calculated (in the same manner as for aluminum). Thus, using Equation 6.1 with a T_0 value of 20°C, a T_{eq} value of 112°C is obtained; that is, the steel wire temperature has increased 92°C in a single pass going from an ambient value to a value greater than the boiling point of water. With poor lubricant heat transfer (dry soap, etc.), tandem steel drawing has the potential to reach

Table 6.2 Values for hypothetical drawing passes for aluminum and for steel

	Ave. strength (MPa)	Finish diameter (cm)	Reduction	Die angle (°)	Friction coefficient	Drawing speed (cm/s)
Aluminum	160	0.050	0.207	8	0.10	2×10^3
Carbon steel	550	0.457	0.324	6	0.05	5×10^2

Note: The die angles given are half-angles.

extremely high temperatures, which encumbers lubrication and brings undesirable metallurgical changes. Temperature control is a major aspect of commercial steel drawing.

Applying the steel data to Equation 6.2 results in an L_{eq} estimate of 34 cm, a value approximately 74 times that of the diameter. This is a much larger value than the one calculated for aluminum due to the large difference in the squared diameter term. However, it is still less than the distance to the drawing capstan, particularly if "dancers" (speed adjustment systems) are used.

6.1.4 The contribution of uniform deformation

The drawing stress and work per unit volume involve contributions of uniform deformation, non-uniform or redundant deformation, and friction as discussed in Section 5.1. It is instructive to examine the contributions of each to the drawing temperature increase in the pass, acknowledging that together they all add up to the last term in Equation 6.1. First, we consider the contribution of uniform deformation to the drawing temperature increase. Equation 5.5 sets forth the uniform work that occurs during a drawing pass. This may simply be divided by $(C\rho)$ to yield the adiabatic drawing temperature increase associated with uniform work, $(T_{uw} - T_0)$; that is,

$$(T_{uw} - T_0) = \sigma_a \ln[1/(1-r)]/(C\rho). \qquad (6.3)$$

6.1.5 The contribution of redundant work

Equation 5.7 sets forth the redundant work that occurs during a drawing pass. This may simply be divided by $(C\rho)$ to yield the adiabatic drawing temperature increase associated with redundant work, $(T_{rw} - T_0)$; that is,

$$(T_{rw} - T_0) = (\Phi - 1)\sigma_a \ln[1/(1-r)]/(C\rho). \qquad (6.4)$$

From Figure 5.2b it is apparent that the redundant work per unit volume increases with radial position. Thus, the level of $(T_{rw} - T_0)$ will be higher at the outer radial positions. The resulting temperature gradient is usually small in relation to the friction heating gradients discussed next, and will be ignored in this analysis.

6.1.6 The total contribution of deformation

Equations 6.3 and 6.4 are simply added to get the total contribution of deformation, $(T_w - T_0)$, to the temperature increase in the drawing pass; that is,

$$(T_w - T_0) = \Phi\sigma_a \ln[1/(1-r)]/(C\rho) \qquad (6.5)$$

6.1.7 The contribution of friction

Frictional heating occurs at the interface between the wire and the die, and thus a radial temperature gradient is established. The gradient is, as noted previously, rather short-lived, and by the time the wire reaches a distance L_{eq}, downstream from the die exit, the contribution of frictional heating will have equilibrated to the frictional work set forth in Equation 5.9. This may simply be divided by $(C\rho)$ to yield the volume-averaged adiabatic drawing temperature increase associated with friction work, $(T_f - T_0)$; that is,

$$(T_f - T_0) = \mu \cot\alpha \; \Phi\sigma_a \ln[1/(1-r)]/(C\rho) \qquad (6.6)$$

It is very important to know the extent of the transient radial thermal gradient; however, since the high temperatures often involved can affect lubrication, wire surface chemistry, and the metallurgical character of the outer portions of the wire. Rigorous evaluation of the frictional heating transient involves complex aspects of heat transfer and data regarding the wire–die interface that are very difficult to develop. Some quantification of the situation can be achieved by estimating the wire *surface* temperature at the die exit, T_{max}, for higher speed drawing. Using the approach of Siebel and Kobitzsch,[4] the expression

$$T_{max} = (1.25)\mu \; \Phi \; \sigma_a[(vL_d)/(C\rho K)]^{1/2} + \Phi\sigma_a \ln[1/(1-r)]/(C\rho) + T_0 \qquad (6.7)$$

can be developed.

A semi-quantitative description has been illustrated in Figure 6.1 wherein T_0, T_w, and the wire surface temperature are plotted versus position along the wire axis from upstream of the die to a downstream position beyond L_{eq}. The line representing T_w is the wire temperature due to starting temperature, T_0, plus the contribution of deformation. This value increases linearly while the wire is traveling through the die. The line representing the wire surface temperature increases to a value of T_{max} at the die exit, and then declines to T_{eq} at the position corresponding to L_{eq}.

6.1.8 Further illustrative calculations

Again, consider a hypothetical aluminum wire drawing pass with the values shown in Table 6.2. With Equation 5.8 it can be shown that Φ is 1.35.

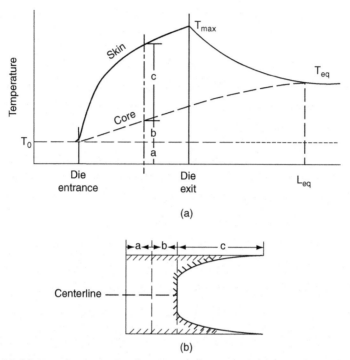

(a)

(b)

Figure 6.1 (a) A semi-quantitative description wherein T_0 (ambient), T_w (core) and the wire surface (skin) temperature are plotted versus position along the wire axis from upstream of the die to a downstream position, beyond T_{eq} and L_{eq}. (b) A plot of the contributions of T_0, T_w, and T_f, as a, b, and c, respectively, as a function of radial position in the drawing channel. Based on E. Siebel and R. Kobitzsch, *Stahl und Eisen*, 63(6) (1943) 110.

Thus, Equation 6.5 results in a temperature increase from deformation work of 21°C, and Equation 6.7 results in a maximum temperature of 102°C + T_0, or 122°C if T_0 is 20°C. As noted in Section 6.1.3, T_{eq} will be 55°C at a distance, L_{eq}, of 0.25 cm downstream from the die. This implies that $(T_f - T_0)$ must be about (55 − 21 − 20) or 14°C, as confirmed by Equation 6.6.

Similarly, again consider a hypothetical steel wire drawing pass with the values shown in Table 6.2. With Equation 5.8 it can be shown that Φ is 1.05. Thus, Equation 6.5 results in a temperature increase from deformation work of 63°C, and Equation 6.7 results in a maximum temperature of 491°C + T_0, or 511°C if T_0 is 20°C. As noted in Section 6.1.3, T_{eq} will be 112°C at a distance, L_{eq}, of 34 cm downstream from the die. This implies that $(T_f - T_0)$ must be about (112 − 63 − 20) or 29°C, as confirmed by Equation 6.6.

6.2. TEMPERATURE MEASUREMENT

The best practical temperature measurements in drawing involve thermocouples. With slow drawing of larger gages, particularly in a laboratory setting, a contact thermocouple can be held against the emerging rod. With higher speeds, lighter gages, and lubricants having poor thermal conductivity, equilibrated drawing temperature can be estimated, or at least correlated, with a contact thermocouple reading at the capstan.

The use of pyrometry is compromised by the rapidly changing emissivity of the wire surface as it leaves the die. This is due to rapid oxidation and other processes of the freshly drawn metal surface. Such rapidly changing emissivity can result in erroneous and confusing pyrometer readings, showing, for example, the appearance of temperature increase rather than decrease downstream from the die.

Die temperatures have been measured, interpolated, extrapolated, and modeled under research laboratory conditions by placing thermocouples at various distances from the die–wire interface. A classic example of such data is shown in Figure 6.2 from the work of Ranger.[28] Ranger's results reflect use of thermocouple placement at the die surface, measurement of drawing

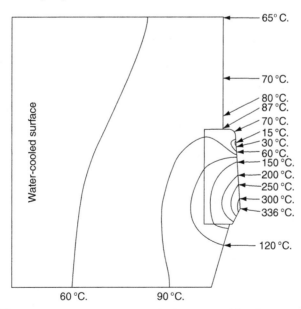

Figure 6.2 Die temperature measurements/estimates resulting from steel drawing at 700 cm/s. From J. G. Wistreich, *Metallurgical Reviews*, 3 (1958), 97, based on A. E. Ranger, *Journal of the Iron and Steel Institute* 185 (1957) 383. Copyright held by Maney Publishing, London, UK, www.maney.co.uk/journals/mr and www.ingentaconnect.com/content/maney/mtlr.

force, and the use of a computational model based on the work of Siebel and Kobitzsch.[4] Steel drawing was undertaken at a speed of about 700 cm/s.

6.3. INTERPASS COOLING

The deformation and friction heating discussed in the previous section can be substantially dissipated between passes. This is especially the case with fine wire drawn with water-based lubricants.

With dry drawing, however, cumulative heating can be substantial, and sophisticated equipment and process design are necessary to maintain drawing temperature at a level consistent with good drawn product quality, stable lubricant performance, and so forth. This is particularly the case with steel drawing. Attention is paid to achieving good heat transfer at the capstan and cooling the dies (where practical). Moreover, drawing schedules may be engineered to keep the heating per pass constant, as discussed in Section 9.6.5.

The small temperature gradients at the die surface in Figure 6.2 indicate the limitations of die cooling with water (heat flow proportional to the gradient). Laboratory drawing experiments indicate that the reduction in peak temperatures through water cooling of dies is only a few percent. While a few percent may be critical in some instances, the opportunities for water cooling are much greater at the capstan, where internal surfaces can be engineered to maximize the cooling effects of internally circulating water.

6.4. PRACTICAL EXAMPLES OF DRAWING TEMPERATURE EFFECTS

6.4.1 Effects on lubricants

The coefficients of friction in drawing with oil-based lubricants directly reflect the viscosities of the lubricants; that is, frictional stresses increase with increasing viscosity in the functional temperature range of the lubricant (for Newtonian viscosity, the shear stress is directly proportional to viscosity). Table 6.3 shows the effects of temperature on viscosity for naphthenic oil at a pressure of 300 MPa.[29,30]

Frictional stresses with solid soap lubrication most directly reflect the shear strengths of the lubricants. Table 6.4 shows the effects of temperature on initial shear strength for sodium stearate.[29,31]

As will be discussed in Chapter 7, there are optimum ranges for lubricant parameters. Thus, increases in temperature during drawing may be

Table 6.3 Relations between temperature and viscosity for naphthenic oil at a pressure of 300 MPa

Temperature(°C)	Viscosity(Pa-s)
38	240
99	0.92
218	0.18

Table 6.4 Relations between temperature and initial shear strength for sodium stearate

Temperature(°C)	Initial Shear Strength(MPa)
80	0.3
99	0.2
119	0.16
142	0.05

beneficial as well as detrimental, depending on the context. The pronounced effect of temperature on lubricant behavior is indisputable. The author has published a practical review of physical conditions in the lubricant layer.[32]

6.4.2 Effects on Recovery and Recrystallization

The temperature increases that occur in commercial non-ferrous drawing can easily bring about recovery and recrystallization during multi-pass drawing. Moreover, these responses further alter the structure-properties response during subsequent annealing.

Figure 6.3 shows the combinations of cumulative drawing reduction and drawing temperature that result in recrystallization in ETP copper wire (from the work of Noseda and Wright[33]). The cumulative true strain of reduction, or ln $[1/(1 - r)]$, is plotted horizontally for as-received, annealed wire and for four subsequent 1 AWG reductions (A, B, C, and D). The drawing temperature is plotted vertically. At least some recrystallized microstructure was observed for conditions to the upper right of the curve labeled 0%. Completely recrystallized structure was observed for conditions to the upper right of the curve labeled 100%. These data were produced by drawing at a speed of 50 cm/s with an oil-based lubricant under laboratory-maintained isothermal drawing conditions. The hardnesses developed by isothermal drawing at different temperatures are represented by dotted

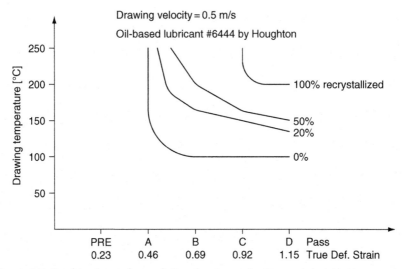

Figure 6.3 Combinations of cumulative drawing reduction and drawing temperature that result in recrystallization in ETP copper wire. From C. Noseda and R. N. Wright, *Wire Journal International*, 35(1) (2002) 74.

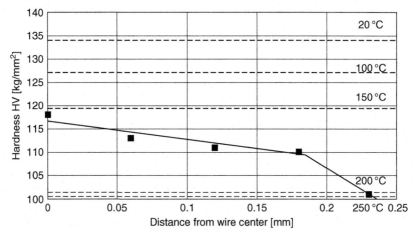

Figure 6.4 Hardnesses developed by isothermal drawing at several temperatures are represented by dotted horizontal lines. Hardnesses measured at radial positions in industrially drawn 0.5 mm ETP copper wire are indicated by the black squares. From C. Noseda and R. N. Wright, *Wire Journal International*, 35(1) (2002) 74.

horizontal lines in Figure 6.4.[33] Hardnesses measured at radial positions in industrially drawn 0.5 mm ETP copper wire are indicated by the black squares in this figure (taken from the work of Wright[34]). These hardnesses suggest that the wire center was generally a little above 150°C with a wire

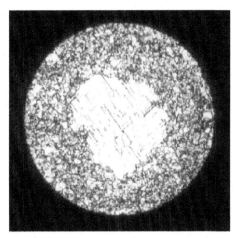

Figure 6.5 Etched cross section of annealed 0.29 mm ETP copper wire, where the outer, frictional-heating-affected regions display a fine-grained recrystallized structure, while the center displays extraordinary grain growth, known as secondary recrystallization. From R. N. Wright, *Metallurgical Transactions*, 7A (1976) 1891.

surface at 200°C or more during the later passes. These temperatures are consistent with the modeling outlined in Section 6.1, and with the presence of substantial recrystallization. Note that the radial hardness gradient shown in Figure 6.4 is opposite to the redundant-work-related hardness gradient displayed in Figure 5.2b. Thus, in Figure 6.4 the thermal gradient hardness effect has dominated any redundant-work-related hardness gradient.

The presence of recrystallization during drawing, followed by subsequent working and so forth, leads to aberrations in the annealing response of the as-drawn wire. Figure 6.5 displays a rather striking example of this, where the outer, frictional-heating-affected regions display a fine-grained recrystallized structure, while the center displays extraordinary grain growth known as secondary recrystallization.[34]

6.4.3 Development of residual stress

As is indicated in Figure 6.1, the wire exits the die with a thermal gradient; that is, with the surface hotter than the center. Moreover, this gradient equilibrates as the wire moves to a distance L_{eq} downstream from the die. During this equilibration process the outer radial portions of the wire (particularly the surface) tend to decrease in length due to the thermal contraction (negative thermal expansion) as these portions cool. To a lesser extent the middle of the wire increases in length as it heats somewhat during equilibration. The overall effect is that the near-surface regions wind up in tension (to be compatible with, or the same length as, the middle regions), and a modest compression develops in the wire center. These stresses, in the as-drawn condition, are called residual stresses, and a great deal of commercially drawn wire contains these stresses.

Residual stresses present a number of problems or points of confusion in as-drawn wire including:

a. Yielding below the nominal yield strength, since stresses below the nominal yield strength, coupled with the residual stresses, can cause yielding
b. The perception of a low elastic modulus, since the yielding cited in (a) will appear at stresses where one normally expects purely elastic behavior (the actual elastic modulus will not have changed, but the slope of the stress-strain curve will be lower due to yielding below the nominal yield strength)
c. Residual stress relaxation, with time, resulting in changes in wire shape, coil set, and so on

Item (a) is particularly apparent in torsion and bending, since the applied stresses are highest at the wire surface, and, in conjunction with residual surface tension, these stresses can cause quite unexpected wire deformation as well as aberrant forming and springback response.

Residual stress can be removed by annealing and stretching where practical.

It should be noted that factors other than frictional heating affect residual stresses in wire drawing, in particular the size of the reduction. For example, very light reductions can produce a residual compressive stress at the surface, thus improving resistance to fracture and fatigue.

The author has published a practical review regarding residual stresses in wire.[35]

6.4.4 Effects on dynamic strain aging in steel

When steel is drawn in the 200 − 600°C range, increased flow stress and as-drawn strength occur due to the dynamic strain aging phenomenon, with higher drawing speeds associated with dynamic strain aging in the upper end of this temperature range. In the lower part of this range, increased drawing speed is associated with *lower* strength. This is opposite of the normal trend, and is sometimes referred to as the anomalous strain rate effect. Actually, process analysis in the dynamic strain aging regime can be quite

Table 6.5 Relations between ultimate tensile strength and reduction per draft for 0.67% carbon steel drawn to a total strain of about 2.4

True Strain (Reduction) per Draft	Ultimate Tensile Strength (MPa)
0.015 (10%)	1950
0.223 (20%)	2020
0.357 (30%)	2240

complicated, and unanticipated results may occur. The author has published a practical review of aging effects in steel wire processing.[36]

An interesting aspect of dynamic strain aging behavior is displayed in Table 6.5 from the work of Dove.[37] This table compares the tensile strength of 0.67% carbon steel drawn to a total true strain of about 2.4 (91% reduction) in three different pass sequences. One sequence involves a strain of 0.105 (10%) per pass, for a total of 23 passes; a second sequence involves a strain of 0.223 (20%) per pass, for a total of 11 passes; and a third sequence involves a strain of 0.357 (30%) per pass, for a total of 7 passes. It is clear that the wire drawn the fewest passes, with the highest reduction per pass, has the highest tensile strength, and that the wire drawn with the most passes, and with the lowest reduction per pass, has the lowest tensile strength. This may be explained by the fact that the wire drawn with the highest reduction per pass experienced the highest drawing temperatures, and thus displays the greatest degree of dynamic strain aging. Again, this is opposite the effect expected from redundant work, which would be lowest in the wire drawn with the highest reductions per pass. Thus, the aging effect has dominated any redundant-work-related effect.

6.4.5 Martensite formation

Martensite is a phase that forms when certain alloys are cooled through and below a critical temperature. Concurrent stress and plastic deformation can affect the martensitic transformation temperatures. Thus the drawing temperature of such alloys can greatly affect drawn properties. This is the case for stock subjected to forming, bending, and torsion, such as in spring and fastener making. Common alloys displaying this sensitivity are austenitic stainless steels and shape memory alloys.

Of course, martensite cannot form unless its parent phase is present, for example, austenite in the case of carbon steel. Carbon steel is rarely (intentionally) drawn in either the austenitic or martensitic condition. However, it is possible for the surface of carbon steel to reach austenite formation temperatures (about 720°C) during drawing, resulting in direct formation of martensite upon cooling and gross embrittlement of the wire.

6.5 QUESTIONS AND PROBLEMS

6.5.1 Using the answers to Problem 5.9.2, calculate the temperature increase for a steel drawing pass, assuming that the temperature is uniformly distributed, and that no heat has been lost to the surroundings. Repeat the calculation for a steel with a tensile strength of 1700 MPa.

Answer: Using Equation 6.1, the temperature increases for steel with an average strength of 350 MPa can be calculated as the drawing stresses divided by the product of specific heat and density (to be found in Table 6.1). The calculated temperature increases are 51.7, 50.9, 51.4, and 55.6 °C. In a similar manner, the calculated temperature increases for steel with an average strength of 1700 MPa are 251, 247, 250, and 270°C.

6.5.2 Verify the L_{eq} values cited in the two illustrative calculations of Section 6.1.3.

Answer: Using Equation 6.2, the L_{eq} value for the aluminum pass is calculated to be 0.251 cm, and for the steel drawing pass it is calculated to be 33.9 cm.

6.5.3 Confirm, from Equation 6.6, the $(T_f - T_0)$ values cited in the illustrative calculation of Section 6.1.8.

Answer: The value of $(T_f - T_0)$ calculated for the aluminum pass is 14.7°C compared to the estimate of 14°C, and the value of $(T_f - T_0)$ calculated for the steel pass is 30.8°C compared to the estimate of 29°C.

6.5.4 Estimate the equilibrated drawing temperatures in the last passes for the three steel drawing schedules cited in Table 6.5. Assume a coefficient of friction of 0.05 and a die semi-angle of 6°.

Answer: Assume that there was sufficient time between passes for the wire to return to an ambient temperature of 20°C. Equation 6.1 allows the temperature increases to be calculated as the drawing stresses divided by the product of specific heat and density (found in Table 6.1). The respective drawing stresses can be calculated from Equation 5.13. The average flow stress can be estimated as the tensile strength, and the respective Δ values for reductions of 0.1, 0.2, and 0.3 are 4.0, 1.9, and 1.2. This leads to drawing stress values of 513, 808, and 1235 MPa, respectively. Finally, the respective temperature *increases* are 143, 224, and 363°C, and the respective equilibrated drawing temperatures are 163, 244, and 383°C.

CHAPTER SEVEN

Drawing Speed

Contents

7.1. DEFINITION AND BASIC FORMULAS

The drawing speed, at any segment of the wire route, is simply the distance a point in the wire travels in a unit time, such as meters per minute. Generally the drawing speed is treated as constant when coming off payoff systems and going onto take-up systems. The speed is also treated as constant between dies. Changes in drawing speed, or accelerations, occur during the drawing pass, since the wire length increases as the cross section decreases. Practical drawing speeds range from 10 to 5000 m/min.

If one assumes that the wire volume does not change, and hence the *product* of cross-sectional area and length does not change, then the product of velocity and cross-sectional area will not change. On this basis,

$$V_0 \times A_0 = V_1 \times A_1, \quad (7.1)$$

or

$$A_0/A_1 = V_1/V_0, \quad (7.2)$$

where V_0 is the wire velocity before the die, and V_1 is the wire velocity upon exiting the die.

Wire Technology
ISBN 978-0-12-382092-1, DOI: 10.1016/B978-0-12-382092-1.00007-5

In the case of bench drawing, the drawing velocity, V_1, is simply the same as the pulling speed. In the case of block drawing or multiple die drawing, V_1 is given by:

$$V_1 = \pi \, D \, \omega, \tag{7.3}$$

where, as in Equation 3.1, D is the drawing block diameter and ω is the block speed in revolutions per unit time.

7.2. THE ROLE OF DRAWING SPEED IN ANALYSIS

7.2.1 Power

As introduced in Section 4.1.4, the power consumed in a given drawing pass can be viewed as drawing force times the exit speed, or $F \times V_1$. If V_1 is in meters per second and F is in newtons, the product is in newton meters per second, which is equivalent to joules per second or to watts.

7.2.2 Strain rate

As introduced Section 4.1.7, the true strain rate of metal deformation, $d\varepsilon_t/dt$, can be important, and in a drawing pass the average strain rate is given by the relation:

$$d\varepsilon_t/dt = \varepsilon_t(V_0 + V_1)/(2L_d), \tag{7.4}$$

where L_d is the length of the deformation zone (see Section 4.2.4). Obviously, the average strain rate increases linearly with the drawing speed. It should be noted that, since strain is dimensionless, the units of strain rate are s^{-1}, and so on.

Large changes in strain rate can have a significant effect on wire flow stress and as-drawn strength, as will be discussed in Chapters 13 and 14.

7.2.3 Drawing temperature

The role of drawing speed in drawing temperature analysis is most obvious during the temperature maximum, at the wire surface as the wire exits the die, as expressed by Equation 6.7:

$$T_{max} = (1.25)\,\mu\,\Phi\,\sigma_a[(vL_d)/(C\rho K)]^{1/2} + \Phi\,\sigma_a\ln\,[1/(1-r)]/(C\rho) + T_0. \tag{6.7}$$

The drawing speed, v, appears in the first of the three terms on the right side of this equation: $(1.25)\, \mu\, \Phi\, \sigma_a\, [(vL_d)/(C\rho K)]^{1/2}$. This term reflects frictional heating. Basically, time is required for the frictional heating (at the surface) to penetrate toward the cooler wire center. Equation 6.7 describes the situation where there is inadequate time for the frictional heat flux to penetrate to the wire center. As the drawing speed increases there is even less time for this flux, and the wire exits the die with an increasingly hotter surface. This contribution is proportional to $v^{1/2}$ for the analysis of Equation 6.7.

Now as the wire emerges from the die, a temperature gradient exists from the hot surface to the cooler centerline, but this will equilibrate a relatively short distance downstream from the die exit. Equation 6.2 describes this distance, L_{eq}, as $(vC\rho d^2)/(24K)$, with such distance proportional to drawing speed. Actually, the equilibration requires a given time, $t_{eq} = L_{eq}/v = (C\rho d^2)/(24K)$.

Once the equilibration of the frictional heating has occurred, one may add Equations 6.5 and 6.6 to T_0 to get the wire temperature, $T_{eq;}$ that is,

$$T_{eq} = (1 + \mu \cot\alpha)\, \Phi\sigma_a\, \ln\, [1/(1-r)]/(C\rho) + T_0, \qquad (7.5)$$

which, in turn, equals $T_0 + \sigma_d/(C\rho)$, as per Equation 6.1.

In any case, Equation 7.5 would seem to include no role for drawing speed. This is not the case, however, because drawing speed affects T_0. In representing the wire temperature at die entry, T_0 includes (in addition to ambient temperature) the effect of heat generation in upstream passes *and* the effect of interpass cooling. Now T_{eq} can be expected to increase with each pass in tandem drawing, since the interpass cooling will generally not remove all of the heat generated in the previous pass, and hence, T_0 will increase.

Typically interpass cooling includes conductive and convective heat loss to air and to lubricant and conductive heat loss upon contact with the capstan. As drawing speed increases, there is less time for such heat transfer and T_0 increases. In high speed drawing, with poorly conducting lubricants, the cumulative increase in T_0 can present serious problems with lubricant stability and performance and with the metallurgical response of the wire (see Chapters 6 and 14).

In summary, *deformation* heating in a given pass does *not* depend on drawing speed unless there is a significant effect of strain rate on flow stress. Similarly, equilibrated frictional heating (downstream) from a given pass does *not* depend on drawing speed. However, interpass cooling opportunities are reduced as drawing speed increases, resulting in a cumulative increase in drawing temperature from pass to pass. And, perhaps most significantly, the surface temperature of the wire at the die exit depends substantially on the drawing speed.

7.3. THE EFFECT OF DRAWING SPEED ON LUBRICATION

As previously noted, increased drawing speed can increase drawing temperature, thus impacting lubricant stability and performance.

Beyond this, drawing speed generally has a direct effect on lubricant film thickness and the related coefficient of friction. This is shown generically by way of the Stribeck curves seen in Figure 7.1. In the lower

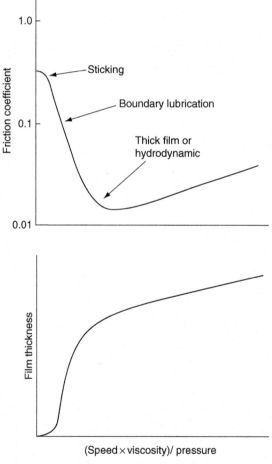

Figure 7.1 Stribeck curves displaying generic dependence of lubricant film thickness and coefficient of friction on drawing speed (as well as on viscosity and die pressure).

portion of the figure, lubricant film thickness increases with the parameter ($\eta v/P$), where η is lubricant viscosity, and v and P are relative interface velocity (drawing speed) and die pressure, respectively. At a certain range of ($\eta v/P$) values, the lubricant film thickness increases grossly, and thus a gross change in lubricant performance at a certain drawing speed range can reasonably be expected.

This increase in lubricant film thickness with speed establishes what is sometimes called hydrodynamic lubrication (liquid or wet lubricants) or simply "thick film" lubrication (dry lubricants). In the upper part of Figure 7.1, it may be seen that the coefficient of friction may drop over an order of magnitude with the establishment of this thick lubricant film.

Finally, with regard to ($\eta v/P$), it should be noted that the increase in temperature associated with increased drawing speed can lower lubricant viscosity (see Chapter 6), and this decrease in η may offset at least some of the effect of a given increase in v, so that the overall increase in ($\eta v/P$) is somewhat reduced.

7.4. SOME PRACTICAL ISSUES

7.4.1 Motivations for higher speed

The primary motivation for increased drawing speed is no doubt that of increased productivity, and with effective temperature control it may be practical to draw at very high speeds (5000 m/min). This is especially the case since lubricant film thickness can be expected to increase with appropriately designed lubricants. And most drawing lubricants are designed with higher commercial speeds in mind.

One drawback to higher drawing speed is the necessity for stable and reliable drawing machine dynamics, particularly regarding take-up systems. In the worst-case scenarios, the advantages of high speed drawing may be completely offset by the requirements for the wire-handling systems.

7.4.2 Start-up concerns

Simply put, upon start-up the wire speed increases from zero, and usually traverses a speed range well below that of the intended steady state. In this range the lubricant may not be adequate, as implied by the relationships of Figure 7.1. The wire may stick to the die and the dies may even be damaged. Such conditions are also pertinent to the very slow drawing speeds involved with die string-up.

Thus, it is common practice to apply a supplemental lubricant to the wire to improve lubrication during start-up and particularly string-up procedures.

7.4.3 Steady-state slow drawing

Bench and heavy gage block drawing are generally undertaken at slow speeds, and lubricant design for such cases should take this into account. Beyond this, some normally high speed drawing operations may be operated in tandem with a slow speed operation and must be operated at reduced speed. For example, copper magnet wire may be drawn in tandem with an enameling operation. In these instances, it is essential that the lubricant selection reflect the requirements of reduced-speed drawing.

7.5. QUESTIONS AND PROBLEMS

7.5.1 A certain drawing pass reduces a wire from a diameter of 1.0 mm to a diameter of 0.9 mm. The value of V_0 is 200 m/min and the block diameter (downstream from the die) is 15 cm. What is the block speed in rpm?
Answer: Using Equations 7.1 and 7.3, $V_1 = V_0(A_0/A_1) = \pi D\omega$ and thus $\omega = V_0(A_0/A_1)/(\pi D)$. Putting in the various values, the value of ω turns out to be 524 min^{-1}.

7.5.2 Using Figure 7.1, approximate the coefficients of friction representative of sticking friction, boundary lubrication, and thick film lubrication. What would be the impact of such friction variations on Equation 7.5?
Answer: Note that the vertical axis in Figure 7.1 is logarithmic. Approximate values of friction coefficient for sticking friction, boundary lubrication, and thick film lubrication are 0.3, 0.1, and 0.02, respectively. Equation 7.5 is $T_{eq} = (1 + \mu \cot\alpha) \Phi\sigma_a \ln [1/(1 - r)]/(C\rho) + T_0$, and the coefficient of friction directly affects the term $\mu \cot\alpha \Phi\sigma_a \ln [1/(1 - r)]/(C\rho)$.

7.5.3 Would a coefficient of friction of zero totally eliminate heating in a drawing pass? Why or why not?
Answer: No, because deformation heating would still remain. Consider Equation 7.5, or $T_{eq} = (1 + \mu \cot\alpha) \Phi\sigma_a \ln [1/(1 - r)]/(C\rho) + T_0$. Even if μ is zero, T_{eq} still equals $\Phi\sigma_a \ln [1/(1 - r)]/(C\rho) + T_0$.

7.5.4 It is common practice to use a different lubricant during string-up and normal drawing. Using the Stribeck curve, explain the basis of this practice.
Answer: Suppose, for example, the lubricant is designed such that $(\eta v/P)$ will be in the thick film range at normal drawing speed. In many cases, (η/P)

will not vary as v varies, so during string-up, when v is very low, the value of (ηv/P) may be in the boundary or even sticking regime, perhaps leading to scratching, and so forth. In principle, the appropriate value of (ηv/P) can be restored by using a provisional lubricant of increased viscosity so that (ηv) can remain about the same.

CHAPTER EIGHT

Friction, Lubrication, and Surface Quality

Contents

8.1. MODES OF LUBRICATION AND RELATED FRICTION RESPONSE

The discussion in Section 7.3 involved the changing of lubricant film thickness as a function of pressure, relative surface speed, and lubricant viscosity. Film thickness is a very practical way to introduce and define lubrication modes.

8.1.1 Little or no film thickness and the sticking friction mode

This extreme of lubrication involves an absence of lubricant between the wire and the die. Given the high die pressure and the continued appearance

Wire Technology
ISBN 978-0-12-382092-1, DOI: 10.1016/B978-0-12-382092-1.00008-7

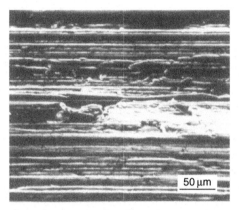

Figure 8.1 Region of an aluminum rod surface drawn under conditions of sticking friction. From R. N. Wright and A. T. Male, ASME Transactions, *Journal for Lubrication Engineering*, Series F, 97(1) (1975) 134.

of unreacted, nascent metal at the wire surface, a relatively strong bond develops between the wire and the die. In this case, the metal sticks to the die, and the shearing action between the wire and the die involves a shearing off of the wire surface, leaving wire metal stuck in the drawing channel. In effect the emerging wire surface is a ductile fracture surface.

In the worst-case scenario, the shear stress between the wire and the die is τ_o, the shear strength of the wire metal. Since the minimum value of the die pressure is σ_o or $2\tau_o$, the maximum value of the coefficient of friction, μ_{max}, is $\tau_o/(2\ \tau_o)$ or 0.5. Actually, it can be argued, on the basis of the von Mises yield criterion and axisymmetric strain, that the theoretical maximum value of μ_{max} is 0.577. Sticking is not likely to occur over the entire wire–die interface, and coefficient of friction values of, say, 0.25 are often used to model sticking friction.

Figure 8.1 displays a region of an aluminum rod surface drawn under conditions of sticking friction.[38] While such conditions will normally lead to a catastrophic breakdown of the drawing operation, certain specialized drawing operations, such as those involved in producing tungsten lamp filaments, can be sustained while generating surfaces such as shown in Figure 8.1.

8.1.2 Local sticking

The gross sticking friction cited in Section 8.1.1 is rather uncommon as far as steady-state drawing is concerned. However, it is often the case that marginal lubrication leads to local sticking, typically leading to long strings of chevron-like shear fractures in the wire surface commonly called "crow's

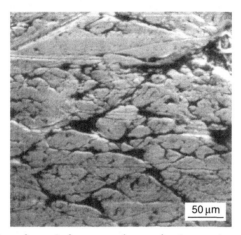

50 μm

Figure 8.2 A region of crow's feet on a drawn aluminum wire surface. From R. N. Wright and A. T. Male, ASME Transactions, *Journal for Lubrication Engineering*, Series F, 97(1) (1975) 134.

feet." As this condition of local sticking becomes more pervasive, involving patches of shear fractures and crow's feet, the condition of sticking friction is approached. Coefficient of friction values of, say, 0.15 are often used to represent conditions of local sticking.

Figure 8.2 displays a region of crow's feet on a drawn aluminum wire surface.[38] As will be discussed in Section 8.8, progressive development of crow's feet can lead to fracture of the wire during drawing.

8.1.3 Boundary lubrication

The concept of boundary lubrication involves the entire wire surface separated from the die by the presence of a minimal (monomolecular?) lubricant film. In principle, there should be no areas of lubricant breakdown, local sticking, or development of crow's feet. On the other hand, the thin film allows a smooth die surface (assuming such exists) to "iron out" or flatten the wire surface, leading to a "bright-drawn" condition.

The ironing process involves considerable plastic deformation of the wire surface, however, and a relatively large coefficient of friction, say 0.10, will exist. Figure 8.3 displays an aluminum rod surface produced under boundary lubrication conditions.[38] It should be noted that a macroscopically bright wire surface can also be produced in the context of local sticking. Thus, a bright wire surface does not necessarily imply purely boundary lubrication, and efforts to achieve bright drawing conditions often run the risk of developing crow's feet.

Figure 8.3 Aluminum rod surface produced under boundary lubrication conditions. From R. N. Wright and A. T. Male, ASME Transactions, *Journal for Lubrication Engineering*, Series F, 97(1) (1975) 134.

8.1.4 Thick film lubrication

Under certain circumstances, the thickness of the lubricant film can become large in comparison to the asperities or "bumps" on the surface of the die and the wire, and such lubrication is called "thick film." Formal tribologists may properly distinguish between films that are, for example, one-half order of magnitude thicker than the asperities, and films that are an order of magnitude or more thicker than the actual asperities. However, for our purposes we will merely consider the category of thick film by which we mean to describe a thick film of lubricant whose intrinsic properties (lubricant shear strength, viscosity, etc.) produce the shear stress between the wire surface and the die.

In Figure 7.1, it is clear that there is a dramatic increase in film thickness at a certain range of ($\eta v/P$). In the case of fluids, such conditions are generally called hydrodynamic film conditions. For a given drawing lubricant, there may be a critical speed range at which hydrodynamic conditions are achieved. Or, for a given drawing speed, a critical viscosity range may be necessary, reflecting the chemical formulation of the lubricant and/or its temperature.

In certain solid lubricants (e.g., solid soap powders) a thick film may be developed, especially if the lubricant pressure at the die entry is enhanced by pressure-enhancing devices (pressure dies, Christopherson tubes, etc.). Such a film is sometimes called quasi-hydrodynamic.

The shear strength of the thick film lubricant, τ_{lub}, is apt to be very low, and the apparent coefficient of friction, μ, or τ_{lub}/P, can be below 0.05 and even as low as 0.01 in some cases. The presence of a thick, rather soft

Figure 8.4 Aluminum rod surface produced under thick film lubrication conditions. From R. N. Wright, *Metal Progress*, 114(3) (1978) 49.

lubricant film eliminates any practical possibility of ironing, and indeed, the wire surface plasticity is sufficiently free to display shear bands and grain-to-grain variations in strain, as shown in Figure 8.4.[39] The macroscopic appearance of such a surface may be dull, or that of a "matte" finish. Such a surface, nonetheless, represents excellent lubrication.

Under conditions intermediate between boundary and thick film lubri-cation, local areas of thick film lubrication, called "thick film pockets," may develop and create local depressions in the worked surface, as suggested by the surface appearance in Figure 8.5.[38]

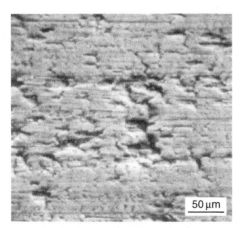

Figure 8.5 Aluminum wire surface drawn under conditions intermediate between boundary and thick-film lubrication. From R. N. Wright and A. T. Male, ASME Transactions, *Journal for Lubrication Engineering*, Series F, 97(1) (1975) 134.

Table 8.1 General relationships of lubricant mode, friction coefficient, and surface appearance

Lubricant Mode	Friction Coefficient	Surface Appearance
Gross sticking friction	0.25?	Heavy striations, cracking
Local sticking	0.15?	"Crow's feet"
Boundary	0.10?	Smooth, bright
Thick film or hydrodynamic	0.03?	Lumpy, dull, matte

8.1.5 Overall comparison

The relationships of friction coefficient, lubrication mode, and surface appearance are summarized in Table 8.1.

8.1.6 The use of a constant value to characterize frictional shear stress

As noted in Section 4.2.6, the use of a coefficient of friction, μ, implies that the frictional shear stress between the wire and the die, τ, is equal to μP. This concept works reasonably well in the boundary lubrication range, where increasing die pressure enhances wire-asperity-to-die-asperity contact, and hence increases τ. However, in the thick film range we have noted that τ is more properly τ_{lub}, and in the sticking friction range τ approaches τ_o, the wire shear strength. Thus, in these ranges it is more appropriate to designate the frictional shear stress as a somewhat constant value, rather than a value that is proportional to the die pressure.

Some mechanical modelers prefer to use a constant value of frictional shear stress between the wire and the die. In particular, a friction factor, m, is utilized, such that $\tau = m \tau_o$, with τ, m, and τ_o all at fixed values.

In determining whether to use a coefficient of friction or a friction factor, τ versus P is plotted. If a linear proportionality is demonstrated, then use of a coefficient of friction is validated. If the value of τ does not increase with P, then the use of a friction factor may be appropriate. Development of such data can be tedious, and an arbitrary decision is often made in opting to use μ or m, based on, say, data availability or mathematical simplicity, including software methodology.

The majority of classical wire drawing analysis employs a coefficient of friction, whatever the physical circumstances. Even so, it is easier to explain certain types of lubricant behavior by way of intrinsic lubricant film strength with a given viscosity.

8.2. PHYSICAL CONDITIONS IN THE LUBRICANT LAYER

8.2.1 The role of viscosity

Given the complexity alluded to in the previous section, it is worth noting the expected responses of the lubricant to factors such as temperature and pressure. In so doing we will focus on the physical property of viscosity. Viscosity is relatable to lubricant shear strength. For example, fluids exhibiting Newtonian viscosity adhere to the relationship:

$$\tau = \eta(d\gamma/dt) = \eta(dv/dy) \tag{8.1}$$

where $(d\gamma/dt)$ is simply the shear stain rate in the lubricant, and (dv/dy) is the velocity gradient across the lubricant thickness. The value of dv/dy is approximately equal to the drawing speed divided by the lubricant film thickness (with the y direction perpendicular to the lubricant film). Of course, not all fluids are Newtonian. However, it is fair to say that the frictional shear stress can be expected to increase with the increasing viscosity of a thick film lubricant. It can also be expected to increase with drawing speed once the thick film is established, consistent with the increase in μ displayed on the right-hand side of the upper portion of Figure 7.1.

Bloor et al.[40] have developed a formula expressing lubricant film thickness at drawing zone entry, y_o, as:

$$y_o = 3\eta v/(\sigma_o \tan\alpha). \tag{8.2}$$

The right-hand side of Equation 8.2 is proportional to $(\eta v/P)$, at drawing zone entry, and $(\eta v/P)$ is the value plotted along the abscissa of Figure 7.1. This is consistent with the linear relationship on the right-hand side of the lower portion of Figure 7.1.

8.2.2 The effect of pressure and temperature on lubricant viscosity

The effect of pressure and temperature on the viscosity of two mineral oils is shown in Figure 8.6, based on the summaries by Schey and Booser.[29,30] It is clear that there is a nearly linear effect of increased pressure on the logarithm of viscosity. A relationship between viscosity and pressure is set forth with the Barus formula:

$$\eta = \eta' \exp(zP), \text{ or } \ln \eta = \ln \eta' + zP, \tag{8.3}$$

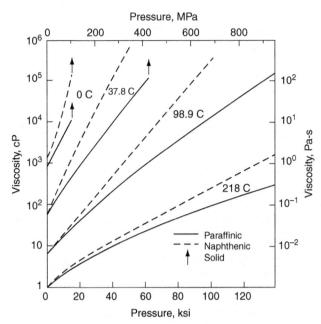

Figure 8.6 The effect of pressure and temperature on the viscosity of two mineral oils. From E. R. Booser, Kirk-Othmer Encyclopedia of Chemical Technology, 3rd Edition, Vol. 14, Wiley-Interscience, NY, 1981, 477. Reprinted with permission of John Wiley & Sons, Inc. Copyright held by John Wiley & Sons, Inc.

where η' is the lubricant viscosity at ambient pressure, and z is a pressure coefficient. Die pressure will vary comparatively little in the course of wire drawing, and the pressure relationships of Figure 8.6 are mostly useful in comparing the response of the lubricant to the drawing of wires of different strengths, where average pressure is proportional to wire flow stress.

Wire surface temperature and lubricant temperature can be expected to vary greatly, however, as discussed in detail in Chapter 6. Thus, it is important to notice in Figure 8.6 that a 100°C increase in lubricant temperature can be associated with viscosity decreases of nearly two orders of magnitude.

Similar declines in viscosity with temperature are displayed in Figure 8.7, again based on the summaries by Schey and Booser.[29,30] Finally, Table 8.2 shows the effect of temperature on viscosity for stearates that are the basis for many solid lubricants, which is from Schey's summary of the work of Pawelski et al.[29,31].

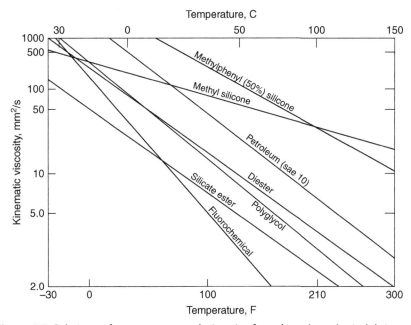

Figure 8.7 Relations of temperature and viscosity for selected synthetic lubricants. From E. R. Booser, Kirk-Othmer Encyclopedia of Chemical Technology, 3rd Edition, Vol. 14, Wiley-Interscience, NY, 1981, 477. Reprinted with permission of John Wiley & Sons, Inc. Copyright held by John Wiley & Sons, Inc.

Table 8.2 Relations of apparent viscosity (in Pa s) to temperature for calcium and sodium stearate and a mixture thereof

Lubricant	75°C	100°C
Calcium stearate	3750	2300
Sodium and calcium stearate	740	280
Sodium stearate	180	150

Detailed analysis of Lucca and Wright[41] has drawn a distinction between wire surface temperature and lubricant temperature, with calculations showing that, under hydrodynamic conditions, the wire surface temperature may actually decrease somewhat with increased drawing speed, concurrent with an increase in lubricant temperature. The basic explanation involves the fact that at higher speeds there is less time for the frictional deformation heating *within the lubricant* to be transferred to the wire or the die. This results in the temperature of the lubricant rising somewhat more than projected in classical assumptions, and actually results in a reduction in the heating of the wire surface as speed increases. The reduction in wire surface heat reflects the decreased viscosity of the warmer lubricant as well as the diminished time for heat transfer.

The practical implication of the Lucca and Wright analysis is that high speed drawing conditions — to the extent that they are represented by hydrodynamic models — result in somewhat more heating of the lubricant than formally thought and in less heating of the wire surface. The argument can be made that the achievement of hydrodynamic lubrication conditions (through high speed drawing or the development of appropriate viscosity for the temperature and pressure conditions) is a major asset for wire drawing, particularly in view of the modest surface heating. Of course, the matte surface implicit in thick film lubrication will be generated.

8.3. QUANTIFYING THE FRICTION STRESS

The frictional stress, and/or the related coefficient of friction or friction factor, is difficult to measure directly. Wistreich[19] and others have employed a split die technique where measurement of the die separating force together with the drawing force allows direct solution of the frictional shear stress. However, this technique generally allows some release of the lubricant through the die split, and establishment of the lubrication mode of a solid die is problematical.

A more practical, if somewhat tedious, approach to measuring the coefficient of friction involves the rearrangement of Equation 5.13 to the form:

$$\mu = (\sigma_d/\sigma_a)[(3.2)/\Delta) + 0.9]^{-1} - \alpha. \qquad (8.4)$$

The value of σ_d can be determined directly from an annular load cell located directly behind the die by dividing the load by the cross-sectional area of the wire exiting the die. The basis for determining σ_a can be Equation 5.14, with the 0.2% offset yield strengths of the entering and exiting wire used for σ_{00} and σ_{01}, respectively. It may be practical to approximate σ_a with the ultimate tensile strength of the wire entering the die. In principle, the value of σ_a should reflect the temperature and strain rate of drawing. Approaches to determining σ_a are discussed in Chapter 11.

Rough estimates of the coefficient of friction can be made by evaluating the surface quality of the entering and exiting wire following the guidelines set forth in Sections 8.1 and 8.6.

8.4. DRAWING WITH HIGH FRICTION

Great effort is expended to keep friction reasonably low in wire drawing by using sophisticated lubricants, coatings, and pressure dies. However, there are products and processes where friction cannot always be low. These include metals with passive layers (such as stainless steels, as discussed in Section 15.2), and certain bright drawing practices. In these cases the coefficient of friction may range from 0.12 to 0.18 and crow's feet may be prevalent. Drawing under these circumstances may be optimized to a certain degree.[42]

A reasonable protocol for high friction drawing involves the following guidelines:
a. Draw stress minimization
b. Avoidance of Δ values much above three
c. Avoidance of σ_d/σ_a ratios above 0.7

Guideline (a) maximizes the range of practical reductions; guideline (b) keeps die pressure, redundant work, and centerline tension within ranges tolerated in many drawing practices; and guideline (c) should keep drawing breaks to a reasonable minimum. Table 8.3 displays some practical drawing values for a reduction of 20%. Reductions at this level should produce the most favorable results for friction coefficients in the 0.12 to 0.18 range.

8.5. REDRAW STOCK SURFACE CONDITIONING ISSUES

Lubrication can be greatly impacted by the conditions of the incoming wire or wire surface, particularly when the wire or "redraw rod" has just been annealed or hot worked. In such cases, vigorous procedures ("pickling," chemical descaling, mechanical descaling, shaving, etc.) may be employed to remove oxide and surface roughness. These procedures may leave surface quality that must be considered when lubrication is subsequently used in the drawing process.

Table 8.3 The range of drawing conditions that appears to be most promising for high friction drawing

Reduction (%)	Friction Coefficient	Included Die Angle (°)	Draw Stress/Flow Stress	Delta
20	0.12	18	0.56	2.8
20	0.15	20	0.62	3.1
20	0.18	22	0.68	3.4

In this context, one should be aware of the following issues:

a. Pickling and chemical descaling procedures, while possibly quite effective, may involve substantial precautions to address environmental and safety regulations

b. Mechanical descaling procedures, such as blasting and surface flexure, may generate particulate matter that must be addressed in terms of environmental and safety regulations

c. Mechanically descaled rod my require different lubrication practice than chemically descaled rod

d. Incomplete surface oxide removal will result in residual oxide that can be quite abrasive, leading to lubricant contamination, scratching, lubricant breakdown, and die wear

e. Rod hot processing schedules that facilitate oxide removal may not optimize mechanical properties

f. Superior surface conditioning may involve combinations of chemical and mechanical descaling

g. Shaving processes generally result in significant scrap and yield loss, and the resulting smooth surface may make lubricant pickup difficult and sticking likely

Redraw stock production may involve the application of surface coatings (conversion coatings) intended to facilitate subsequent lubrication. The coating may increase the adherence of the lubricant to the workpiece surface, and it may actually function as a lubricant. Moreover, it may protect the rod from oxidation and so forth during shipment to the drawing site. On the other hand, the coating may complicate subsequent cleaning of the drawn product. Some examples of practical coatings are

• "Sull coat" (hydrated oxide) plus lime (calcium oxide) for steel drawing
• Oxides, in general, such as in tungsten drawing
• Borax (hydrated sodium borate) for steel drawing
• Phosphates for steel drawing and oxalates for stainless steel drawing
• Soft ductile metals, such as copper on steel
• Polymers

8.6. CHARACTERIZATION WITH MICROSCOPY

Figures 8.1 through 8.5 display scanning electron micrographs; the scanning electron microscope (SEM) is the most facile instrument with which to visualize the wire surface. This is due to its great depth-of-field and wide magnification range. The author has published several papers involving SEM characterization of metalworked surface quality.[38,39,43,44]

One may wish to clean the wire surface with a solvent prior to examination. However, unclean surfaces often display residual lubricant and pressed-in "dirt" and fines that may be useful in the drawing analysis. These surface residues can often be identified by energy dispersive X-ray analysis, among other micro-analytical methods. While details may be instructive at a wide range of magnifications, a magnification of 400× is recommended for analyses concerned with the more common elements of drawing surface quality (inclusions, crow's feet, grain size, striation spacing, etc.).

Visible light microscopy can be useful in characterizing surface quality, particularly with the use of binocular microscopes in the 25 to 100× magnification range. Such microscopy is particularly useful once more definitive views of the surfaces have been established with SEM. Reference atlases of generic surface views can be very helpful for prompt categorization of the observed surface quality.

The major limitation of visible light microscopy for wire surface analysis is depth-of-field, especially given the curved surface to be examined. The surface can be represented in a planar manner with the use of replicas in which a plastic replica is made of the surface. The curved replica can be flattened and perhaps "shadowed" with chromium and so forth to provide a high resolution mirror image of the wire surface. A detailed description of replica technique has been published by Baker and Wright.[45]

8.7. ILLUSTRATIONS OF MICROSCOPIC CHARACTERIZATION AND ANALYSIS

Figures 8.1 through 8.5 have already presented microscopic characterizations that indicate some basic lubrication modes. Figure 8.8 displays a surface quality that is often inherited with rolled redraw rod.[38] The rolling process often involves sticking friction, and as the roll surface moves away from the rod surface, "tongues" of rod surface metal are pulled away by the departing roll surface. Such surface tears, together with striations similar to those shown in Figure 8.1, are displayed in Figure 8.8. The surface features displayed in Figure 8.8 will generally be pressed into the wire surface in the drawing process, as seen in Figure 8.9. If this is unacceptable, the rod surface may be shaved, resulting in an as-drawn surface such as illustrated in Figure 8.10.

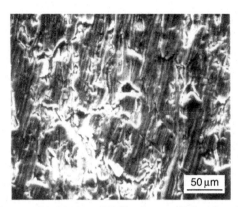

Figure 8.8 SEM micrograph of 1.62 cm EC Al redraw rod, showing the result of sticking friction during hot rolling. From R. N. Wright and A. T. Male, ASME Transactions, *Journal for Lubrication Engineering*, Series F, 97(1) (1975) 134.

Figure 8.9 SEM micrograph of 0.65 cm EC Al rod drawn *without* shaving. From R. N. Wright and A. T. Male, ASME Transactions, *Journal for Lubrication Engineering*, Series F, 97(1) (1975) 134.

Shaving can be an expensive process, especially in view of the amount of scrap generated by removing, for example, 0.25 mm from the rod diameter. Therefore, it may be practical to allow the rolled-rod surface flaws to open up and be ironed out during subsequent drawing. Wright and Male have illustrated such development, starting with hardness indentations as surface flaws.[38] Figure 8.11 shows the initial indentations, while Figures 8.12 and 8.13 show the evolution of the indentations during drawing with calcium-stearate-base and compounded mineral oil lubricants, respectively. With

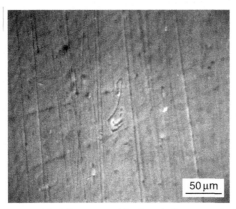

Figure 8.10 SEM micrograph of 0.93 cm EC Al rod drawn *with* shaving. From R. N. Wright and A. T. Male, ASME Transactions, *Journal for Lubrication Engineering*, Series F, 97(1) (1975) 134.

Figure 8.11 Rockwell 15T superficial hardness indentations in 0.74 cm EC Al rods preparatory to drawing. From R. N. Wright and A. T. Male, ASME Transactions, *Journal for Lubrication Engineering*, Series F, 97(1) (1975) 134.

measuring microscopy, it is possible to describe this evolution by way of indentation volume change, as shown in Figure 8.14. Note that the higher strength calcium-stearate-base, thick-film lubricant preserved the indentations somewhat longer than the low shear strength compounded mineral oil boundary lubricant.

In many drawing systems surface quality characterizations reveal damage from contact with surfaces outside of the die, such as rough capstans, crossed-over wires, worn sheaves, and so on. Figure 8.15 displays an otherwise smooth wire surface that has sustained many such blemishes.[38]

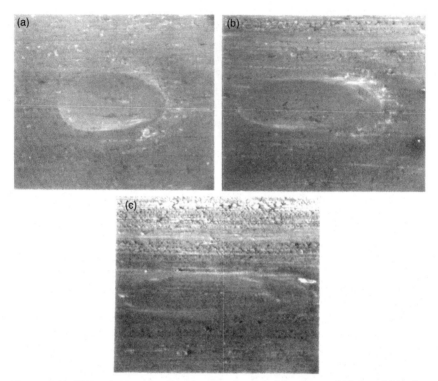

Figure 8.12 SEM micrographs showing hardness indentations of Figure 8.11 after drawing in calcium-stearate-base lubricant (from left to right) to diameters of 0.66 cm (a), 0.57 cm (b), and 0.50 cm (c). From R. N. Wright and A. T. Male, ASME Transactions, *Journal for Lubrication Engineering*, Series F, 97(1) (1975) 134.

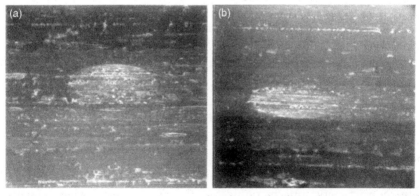

Figure 8.13 Similar to Figure 8.12, except that drawing has been done in compounded mineral oil to diameters of 0.66 cm (a) and 0.57 cm (b). From R. N. Wright and A. T. Male, ASME Transactions, *Journal for Lubrication Engineering*, Series F, 97(1) (1975) 134.

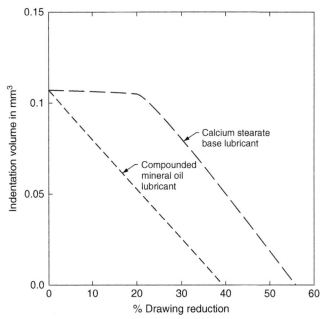

Figure 8.14 Volume of hardness indentations shown in previous figures as a function of drawing reduction. From R. N. Wright and A. T. Male, ASME Transactions, *Journal for Lubrication Engineering*, Series F, 97(1) (1975) 134.

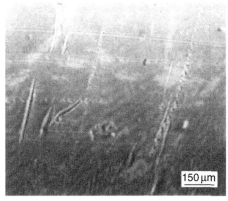

Figure 8.15 SEM micrograph of EC Al wire showing contact scoring. From R. N. Wright and A. T. Male, ASME Transactions, *Journal for Lubrication Engineering*, Series F, 97(1) (1975) 134.

8.8. THE DEVELOPMENT OF CHEVRONS (CROW'S FEET)

8.8.1 General issues

Figure 8.2 displays an area of chevrons, or crow's feet, on an aluminum wire surface after drawing with lubrication at the transition between boundary mode and sticking friction. Actually, the extent of chevron development in Figure 8.2 is rather extensive, suggesting that the causative lubrication conditions have been present at earlier drawing passes. Figure 8.16 shows a schematic illustration of progressive chevron development leading to a drawing break. The first indication of crow's feet (top portion of Figure 8.16) represents local sticking friction, which has lead to a string of chevron-like shear fractures in the wire surface.

Figure 8.17 displays SEM micrographs of three levels of chevrons, together with a reference view of an essentially chevron-free surface.[38] Chevron levels have been classified as follows:

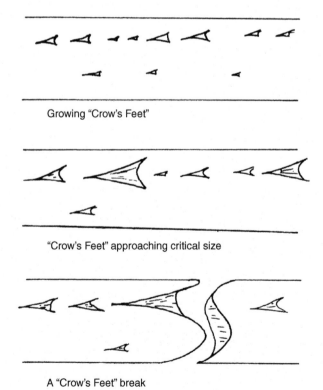

Growing "Crow's Feet"

"Crow's Feet" approaching critical size

A "Crow's Feet" break

Figure 8.16 Schematic representation of progressive chevron, or "crow's feet" development, leading to a drawing break.

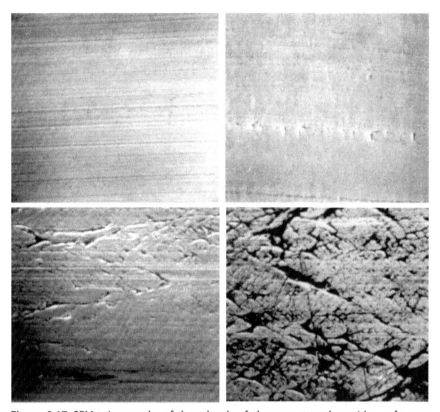

Figure 8.17 SEM micrographs of three levels of chevrons, together with a reference view of an essentially chevron-free surface in the upper left. From R. N. Wright and A. T. Male, ASME Transactions, *Journal for Lubrication Engineering*, Series F, 97(1) (1975) 134.

- Occasional and isolated
- Occasional and sequential
- Frequent
- Frequent and sequential
- Large in relation to wire diameter
- Wire break associated

8.8.2 Mechanics of chevron development[46]

Chevrons can develop in the drawing of any wire when a local breakdown occurs in the boundary lubrication mode and the wire sticks to the die. The portion of the die circumference involved is apt to be small, otherwise gross sticking friction would develop. Once conditions for sticking develop at such a point on the die surface, they may persist for some length of wire drawing, leading to a string of chevrons, as shown in the top part of Figure 8.16.

The local sticking may be "statistical," simply reflecting the random breakdown of marginal or deteriorating lubrication. On the other hand, there is certainly evidence that chevrons develop in conjunction with surface inclusions or contamination that makes lubrication locally difficult.

To develop an understanding of chevron cracking, it is useful to consider the mechanics of the shearing that develops local to the point of sticking. The V-shaped surface fracture mechanics are likely consistent with the strain conditions illustrated in Figure 8.18. In principle, general drawing strains should be added to the strains of Figure 8.18, but these will be ignored because they are rather minor in relation to the local stick–slip mechanics. Figure 8.18 schematically represents an area of wire surface with the drawing direction running left to right. If axes r, θ, and z represent the radial direction, circumferential direction, and axial direction, respectively, in cylindrical coordinates, then Figure 8.18 displays a θ-z surface with wire movement and drawing direction proceeding in the z direction. At the

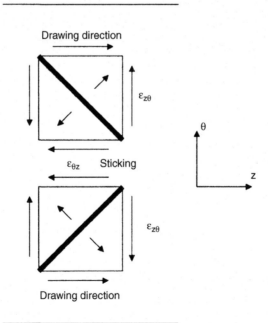

Figure 8.18 Strain conditions consistent with V-shaped surface fracture. From R. N. Wright, *Wire Journal International*, 35(8) (2002) 86.

point of sticking there will be a shear strain, $\varepsilon_{\theta z}$, in the opposite direction of wire movement. Such a strain implies the existence of a conjugate strain, $\varepsilon_{z\theta}$, and the shear couples or θ–z strain states are diagrammed in Figure 8.18. The θ–z surface maximum tensile strain directions are illustrated by the diagonal arrows, and the coarse diagonal lines represent the orientation of a crack that would be developed perpendicular to the direction of maximum tensile surface strain. The coarse diagonal lines are in an opposite direction on each side of the sticking point, and fracture along these lines creates a chevron or V-shaped surface crack.

The fracture path in the r–z plane (at and just below the wire surface) at the point of sticking is indicated by the mechanics illustrated in Figure 8.19. The wire surface is represented at the top of the figure with wire movement and drawing direction proceeding in the z direction. Sticking is creating a shear strain, ε_{rz}, in the opposite direction of wire movement. Such a strain implies the existence of a conjugate strain, ε_{zr}, and the shear couple or r–z strain state is diagrammed in Figure 8.19. The r–z plane maximum tensile strain direction is illustrated by the diagonal arrows, and the coarse diagonal line represents the orientation of a crack that would be developed perpendicular to the direction of maximum tensile strain. Such a crack, emanating from the point of the chevron, would create the undercutting surface easily perceived in Figure 8.20.[47]

As noted previously, the overall strains of drawing must be added to these mechanics for a complete model. The overall drawing mechanics are probably most important to the evolution and growth of the chevrons in subsequent passes. Clearly the cracks enlarge, most likely upon exit from the die.

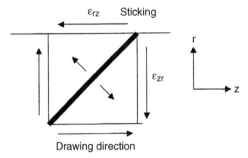

Figure 8.19 Strain conditions consistent with fracture at and below the wire surface. From R. N. Wright, *Wire Journal International*, 35(8) (2002) 86.

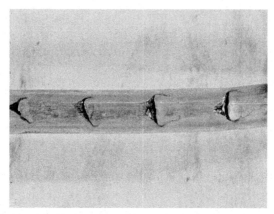

Figure 8.20 Sequential chevrons undercutting a wire surface. (Courtesy of Horace Pops)

The sequential appearance of chevrons in a longitudinal string down the wire surface suggests the persistence of the sticking, together with a momentary, intermittent release of local surface frictional stress, as a fracture occurs. Such stress must build up again before a new chevron can be nucleated. Similar phenomenology may be seen in the case of center bursts, where centerline fractures occur repetitively, as illustrated in Figure 5.7.

8.8.3 Engineering significance of wire surface quality

The potential for chevrons to lead to wire breaks has been amply emphasized. Other risks from chevrons are outlined as follows:

- Can become "cold shuts" or laps entrapping debris and lubricant, presenting the potential for reopening in subsequent mechanical testing, downstream processing, and service manipulations. This is a particularly serious issue with wire that is to be enameled, with welding wire, and with other products where cleanliness is imperative.
- May cause coatings of all sorts to rupture, particularly during wire bending.
- Can propagate to a fatigue failure in wire subjected to cyclic loading in applications such as mechanical cables, brushes, and general reinforcement.
- May propagate to overall fracture in various transient or monotonic loading situations if the wire is sufficiently brittle.

- May exacerbate wire corrosion.
- May simply be objectionable to the customer.

Chevron population and nature can be an important measure of drawing process stability. Increases in chevron population may indicate the following:

- Deterioration in lubricant performance and/or development of lubricant contamination with fines, and so forth
- Emergence of lubrication-complicating factors such as die wear, die misalignment, and surface oxide
- Inheritance of damaged or chevron-containing redraw stock

The previous remarks have centered on the role of chevrons, and chevrons are arguably the most troublesome wire surface flaw that can be directly related to the drawing process. Nonetheless, surface cracks and irregularities present some general problems, whatever their specific cause. Mechanical properties are nearly always compromised by surface cracks. Tensile elongation and fracture stress may be reduced. A particularly large loss of high-cycle fatigue life can be expected, since the fraction of the fatigue life that is spent "nucleating" a crack has been preempted. Surface cracks can also entrap lubricant, and so forth and make cleaning difficult. Surface roughness, in general, can compromise the integrity of enamels and other coatings. Figure 8.21 shows a micrograph of an enameled magnet wire where enamel cracking is obviously related to wire surface asperities (the enamel had been "overbaked" to increase the number of fracture sites).[48]

Over-baked formvar

EC aluminum

Figure 8.21 Micrograph of enameled magnet wire, where enamel cracking is related to wire surface asperities. From R. N. Wright, *Insulation/Circuits*, 20 (13) (1974), 30.

8.9. QUESTIONS AND PROBLEMS

8.9.1 A certain friction study is undertaken resulting in a plot of friction coefficient as a function of die pressure. The friction coefficient is seen to decrease as pressure increases and is inversely proportional. This seems counterintuitive and requires explanation. Provide such an explanation.

Answer: The lubricant is manifesting a constant level of shear strength, τ_{lub}. If a coefficient of friction is used to characterize this friction, then $\mu = \tau_{lub}/P$, and P and μ are inversely related. If τ_{lub} appears to be constant, it should be used to characterize friction; if μ appears to be constant, then it should be used to characterize friction.

8.9.2 Consider the data in Table 6.3. Suppose that wire is drawn with a flow stress of 250 MPa through a die of 12° included angle at a velocity of 10 m/s. If the average die pressure is 300 MPa and if naphthenic oil is the lubricant, what will be the lubricant film thicknesses at 38, 99, and 218°C?

Answer: Use Equation 8.2 with Table 6.3 as a source of the appropriate viscosities, and remember that the included die angle is twice the die semi-angle. The values of lubricant film thickness are projected to be 0.257 mm, 1.05 μm, and 0.021 μm for the temperatures of 38, 99, and 218°C, respectively.

8.9.3 What is the value of lubricant shear strength projected for the three cases in Problem 8.9.2?

Answer: Use Equation 8.1 and approximate (dv/dy) by the drawing velocity divided by the lubricant film thickness. The values of (dv/dy) are 3.6×10^4, 9.5×10^6, and 4.9×10^8 s^{-1}, for respective film thicknesses of 0.257 mm, 1.05 μm, and 0.021 μm. The lubricant shear strength for these cases is 8.8 MPa.

8.9.4 A 1.00 cm diameter rod is shaved to remove the surface to a depth of 0.1 mm. What is the material yield loss? What factors might affect the actual loss of money?

Answer: The cross-sectional area of the rod before shaving is 78.5 mm^2, and the cross-sectional area after shaving is 75.4 mm^2. Thus, the volumetric loss from shaving is about 4%. This loss may be offset by the value of the shavings as scrap, which will be a function of internal recycling opportunities, regional scrap markets, cleanliness issues, and so on.

Drawing Die and Pass Schedule Design

Contents

Wire Technology
ISBN 978-0-12-382092-1, DOI: 10.1016/B978-0-12-382092-1.00009-9

9.1. GENERAL ASPECTS AND THE ROLE OF Δ

9.1.1 Overall geometry

The shape of the deformation zone is by far the most important aspect of wire drawing die design. As discussed in Section 4.2.4, deformation zone shape is characterized by Δ, and Δ is a function of *reduction* and *die angle*. The die angle is the only one of these two variables represented in the die, and it is the most important feature of die design. However, overall die features should reflect the intended reduction, for example, the initial contact should be approximately in the middle of the drawing cone.

The cross-sectional view of a typical carbide wire die is displayed in Figure 9.1. The die involves a steel-encased carbide **nib**, upon which the wire–die contact occurs. Generally, the placement of the nib in the steel case involves a "press fit," where the nib is placed in compression to reduce potentially

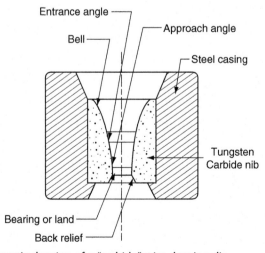

Figure 9.1 Schematic drawing of a "carbide" wire drawing die.

damaging tensile stresses that occur in the die during drawing. The geometry involved in drawing analysis is confined to the nib. The drawing cone is the section of the nib defined by the approach angle. The **approach angle** is generally described as the included, or total angle, whereas α, in the analyses of this book, refers to half this angle. The flared profile upstream from the drawing cone is referred to as the **bell**, although some die designs specify a transitional geometry between the bell and the approach angle, such as the **entrance angle** to be noted in Figure 9.1. Downstream of the drawing cone is a straight cylindrical section referred to as the **bearing** or **land**. The final segment of the die channel involves a chamfer, and is described as **back relief**. Detailed discussion of these geometrical features is presented in this chapter, along with a discussion of die materials. First, however, the impact of the die deformation zone geometry on major process phenomena will be presented.

9.1.2 Effects on friction and heating

For a constant coefficient of friction, low Δ drawing die design is associated with increased frictional heating. The frictional heat is concentrated at the die–wire interface and *may* lead to diminished lubrication, further frictional heating, and catastrophic lubrication breakdown. Concomitant problems are poor wire surface quality and metallurgical changes near the wire surface. There are numerous reasons to use low Δ die designs; however, use of low Δ die designs can only be undertaken in conjunction with stable, relatively low-friction lubrication. This requirement is eased in many cases by the tendency for low approach angles to foster thick film lubrication and a reduced coefficient of friction.

9.1.3 Effects on die pressure and die wear

If stable lubrication can be maintained, low Δ die design should, in principle, provide increased die life. The profile of a typically worn carbide die is displayed in Figure 5.5. A ring of concentrated wear is observed where the wire initially contacts the die together with a more general loss of material throughout the drawing channel. The general wear is a function of the die pressure, P, and the increase in the average die pressure with Δ is given in Equation 5.18 as:

$$P/\sigma_a = \Delta/4 + 0.6. \tag{5.18}$$

A more detailed relationship is displayed in Figure 5.4.

It must be emphasized that *small* reductions (increased Δ) are associated with increased die pressure. The phenomenon is well documented, however, and increased die wear is observed with light passes. The concentration of wear at the locus of initial die contact probably reflects fatigue-like damage associated with the cyclic pressures resulting from variations in entering wire cross section, vibrations of the entering wire, and variations in lubricant flow. Whatever the cause, the severity of the wear ring is observed to increase with increased approach angle (increased Δ). Theoretically, minimum die pressures are not approached unless Δ is lowered to approximately 1.0, or well below the values cited for minimum draw stress, as seen in Equation 5.16. In any case, if lubrication is adequate, die design optimized for good die life will likely involve a Δ value lower than that in Equation 5.16.

9.1.4 Effects on annealing requirements

Another favorable aspect of low Δ die design is that it fosters relatively uniform metal flow with reduced redundant work. Redundant work is plastic deformation over and above that implied by the cross-sectional area reduction, and is usually expressed by the redundant work factor, Φ, or the ratio of total deformation work to the deformation work implied by dimensional change. The redundant work factor is described in Equations 5.6, 5.7, and 5.8, with the relation of Φ to Δ given by:

$$\Phi \approx 0.9 + \Delta/(4.4). \tag{5.8}$$

The redundant work contributes extra strain hardening, particularly at the wire surface, and limits the number of passes and overall reduction that can be taken before annealing is necessary. Thus, low Δ die design should be consistent with reduced requirements for intermediate annealing.

9.1.5 "Central bursts" and finished product ductility

Yet another manifestation of low Δ die design and uniform metal flow is the absence or reduction of hydrostatic tension at the wire center. Such hydrostatic tensile stress has been shown to lead to porosity and central bursts at the wire center. These defects lead, in turn, to "cuppy core" type failures during drawing and to reduced ductility in the as-drawn product, even after *subsequent* annealing. The level of centerline hydrostatic tension is plotted versus Δ in Figure 5.6. Note that substantial centerline hydrostatic tension exists for values of Δ associated with minimum drawing stress (as per Equation 5.16). Thus, yet another argument exists for designing with low Δ values.

9.1.6 Bulging and thinning

Figure 5.6 displays a calculated relationship of Δ to the bulging phenomenon; namely the backing up or upsetting of wire at the die entry. Such behavior is shown literally in Figure 5.8. Bulging undoubtedly increases redundant work and limits the access of lubricant to the drawing cone.

The bulging behavior is accompanied in Figure 5.8 by the phenomenon of wire thinning beyond the die exit. Such steady-state thinning is not to be confused with necking of the wire due to excessive drawing stress or with any sort of elastic behavior (although these factors together with lubricant blockage and die pickup may be additional causes of reduced as-drawn diameter). Back tension (Section 5.8) may also increase the tendency for thinning.

Both bulging and thinning are evidence of metal flow imposed by the die outside the die confines. Both bulging and thinning are observed to increase with increasing Δ, and while the Δ values for bulging in Figure 5.6 are well beyond the range of greatest practical interest, Wistreich observed thinning at Δ values even below those usually associated with minimum drawing stress.[19] Fortunately the thinning behavior seems to be consistent, and does not necessarily imply variability in the as-drawn diameter. Even so, the thinning and bulging phenomena can be used as a further argument for low Δ design.

9.1.7 Summary remarks on the role of Δ

The previous remarks indicate that, provided adequate lubrication can be maintained, die designs involving low Δ values (i.e., low approach angles and/or large reductions) should offer superior performance in terms of reduced wear, reduced requirements for intermediate annealing, reduced cuppy core breakage, improved final product ductility, and minimization of thinning beyond the die exit.

With a few exceptions, continued improvement in performance could probably be achieved by reducing Δ to values as low as one. The value of Δ normally associated with minimum drawing stress is more likely to be in the range of 2 to 3, however (see Section 5.4). Since it is not advisable to have draw stress levels much above six-tenths of the wire-breaking stress, some compromise must be made, and the most practical Δ levels are not likely to be as low as one. The point is that absolute adherence to draw stress minimization will likely lead to Δ values above those leading to optimum engineering performance.

9.2. COMMON DIE MATERIALS

9.2.1 General requirements

A die material must be hard enough to have an acceptable wear rate and tough enough to resist fracture in the face of impact loading and thermal shock. Moreover, it must be economical in terms of the basic service value of the material and the cost of die manufacture.

9.2.2 Historic die materials

As noted in Chapter 2, ancient drawing practices are thought to have involved dies made from natural stones, bored with wooden sticks and abrasive media such as sand/tallow. In late medieval times, drawing plates of hard metals were drilled and opened appropriate to a series of drawing reductions, as suggested by Figure 2.1. With the coming of the industrial revolution, cast irons, high carbon steels, and "tool steels" became available on a practical basis, and certain tool steels are still employed for occasional, limited use.

However, nearly all current focus on drawing die nib materials involves composites of tungsten carbide and cobalt ("carbides") and single crystal diamonds, both natural and synthetic, as well as synthetic polycrystalline diamonds.

9.2.3 Tool steels

Tool steel dies are rarely used in the bulk of the wire drawing industry. However, they can be attractive for certain shaped rod drawing applications and in developmental drawing studies where several iterations of design may be anticipated. Where appropriate facilities exist, tool steel dies are easily made. In any event, a brief consideration of tool steels is a useful precursor to a discussion of carbides.

The term tool steel is used to describe a very wide range of alloy and high alloy steel compositions. Whether one is referring to a high-carbon plain carbon steel or a highly alloyed "high speed" steel, the basic microstructure of these steels in service consists of a large volume fraction of *hard carbide* in a *soft matrix* of iron or iron alloy. To a first approximation, the hard carbide provides wear resistance, while the matrix phase provides some measure of toughness. With plain carbon steels, the carbide phase is iron carbide. Highly alloyed tool steels can contain substantial additions of chromium, molybdenum, tungsten, or vanadium, which form carbides. These carbides are more stable at high temperatures (and high workpiece-tool interface speeds), and resist tool softening and tempering. The hardness of tool steels, in the service condition, can range as high as 700–800 on the Vickers scale, or between quartz (7) and topaz (8) on the Mohs scale.

9.2.4 Composites of tungsten carbide and cobalt

Remarkably effective tool materials can be made by processing tungsten carbide (as well as other carbide) powder with cobalt powder to form a structure with a high volume percent of tungsten carbide bonded with a much smaller amount of cobalt matrix. The resulting bulk material is generically referred to as carbide, although it is really a composite material. Similar composites are sometimes called cermets (ceramic-metal composites). The roles of the hard tungsten carbide and the soft cobalt matrix are similar to the roles of the carbide and the matrix in tool steels. The size of the tungsten carbide particles, or grains, is in the range of a few micrometers, and the percent cobalt can range from a few percent to as much as 25 percent by weight. Hardnesses range from 600 to 1000 on the Vickers scale, although the hardness of the tungsten carbide particles is more in the range of 2000 on the Vickers scale or just above corundum (9) on the Mohs scale.

Composites of tungsten carbide (WC) and cobalt (Co) were developed in Germany early in the twentieth century. They provided hardness, wear resistance, and toughness intermediate between tool steels and diamonds. The composite material also provides good thermal shock resistance. The processing of these composites involves compaction of tungsten carbide, cobalt, and a binder, such as paraffin. The compact is heated to carefully drive off the binder and sinter the tungsten carbide and cobalt at, for example, 1400°C. A cobalt-rich liquid phase is present during sintering.

The resulting microstructure is illustrated in Figure 9.2.[49] The gray, angular phase is tungsten carbide, whereas the lighter phase is cobalt. While this material is generically referred to as carbide, it is composed of the two phases, WC and Co. The role of the Co is to provide toughness, and the toughness of the carbide generally increases with Co content. The hardness and wear resistance generally decrease with Co content, however, as shown in Figure 9.3.[49] Figure 9.3 also shows that hardness and wear resistance generally decrease with increased carbide particle, or grain, size.

Since wire dies involve relatively small carbide components imbedded in relatively tough steel, carbide wire die material typically involves lower cobalt content, say, in the six weight percent range. It should be noted that the density of WC and Co are rather different, namely 17.15 g/cm^3

Figure 9.2 Micrograph of a composite of tungsten carbide (gray phase) and cobalt (white phase). Courtesy of Carboloy Systems Department, General Electric Company.

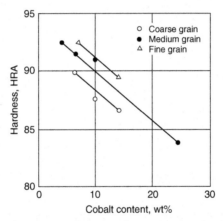

Figure 9.3 Hardness of WC-Co as a function of Co content and carbide grain size. From A. T. Santhanam, P. Tierney and J. L. Hunt, *Metals Handbook, 10th Edition, Vol. 2,* ASM International, Materials Park, OH, 1990, p. 950. Copyright held by ASM International, Materials Park, OH.

and 8.9 g/cm^3, respectively, and thus a six weight percent Co content corresponds to eleven volume percent.

9.2.5 Diamond

Die wear can be reduced to very low levels with the use of diamond die materials. Where a meaningful comparison can be made, die life is often two orders of magnitude higher than for the case of cemented carbide. On the other hand, diamond material is relatively expensive, and sophisticated technique is required for die manufacture.

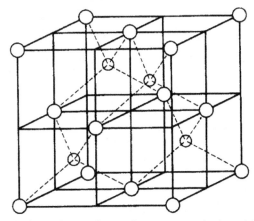

Figure 9.4 Unit cell of the diamond crystal structure, with dotted lines representing close packed directions.

Diamond materials are essentially elemental carbon with an extremely strong chemical bond and a hard crystal structure. Hardness is in the range of 8000 on the Vickers scale, and diamond is the reference material for the top (10) of the Mohs scale. Figure 9.4 displays a "unit cell" of the crystal structure with the carbon atom positions indicated by circles. The circle diameters are smaller than the atoms would be, and the dotted lines in the figure represent the "close packed" directions in which the carbon atoms are in contact.

The diamond structure is not the common state for carbon (which is graphite), and very high pressures and temperatures are required for its formation. Such conditions have developed naturally, and diamonds are mined as a result. Large numbers of natural "industrial" diamonds are in use for mechanical applications. These industrial diamonds often have colors that are undesirable for gem stones, but are otherwise excellent crystalline structures. In the latter part of the twentieth century, diamond synthesizing processes were developed in a laboratory and soon transferred to industrial technologies. Today we have both natural and synthetic diamonds available for wire die applications.

As a drawing die material, diamond offers extremely high wear resistance; however, it is brittle. It also appears to react with carbide forming elements such as iron, thus limiting its use in ferrous drawing. Nonetheless, it is widely used in non-ferrous drawing.

Natural diamonds are single crystals that can be finished to a very smooth surface, commensurate with smooth wire finishes and minimal fine development. However, they wear unevenly due to the directional properties of the crystal structure. Single crystal diamonds can be produced synthetically as well. It should be noted that the orientation of single crystal diamond stones is determined by X-ray diffraction so that the stones can be oriented in the optimum direction as nibs.

Polycrystalline diamonds are produced synthetically. The polycrystalline structures wear more uniformly than their single crystal counterparts and are somewhat tougher. However, polishing can be difficult, surfaces are less smooth, and fines are more readily generated. Polycrystalline diamond die performance seems to be grain-size dependent, although industrial observations vary. Coarser grained structures appear to have a number of advantages, such as superior heat transmission, toughness, lubricity, galling resistance, and fine generation resistance. Finer grained dies are felt to offer superior abrasion resistance. A major factor appears to be the role of "cusp-like" eruptions of the grain boundaries on the polished die surface, with coarse-grained structures having fewer cusps. Corbin published a general discussion of the relative performances of polycrystalline diamond dies.[50]

9.2.6 Effects on lubrication

Baker conducted extensive studies of the effect of die materials on friction in copper wire drawing, and representative data from his work are shown in Figure 9.5 for the drawing of ETP copper.[51] In the case of diamond dies, a

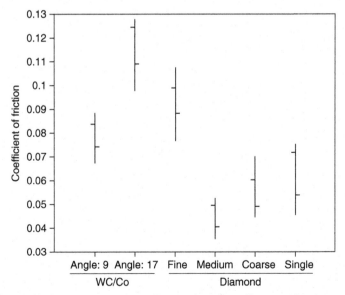

Figure 9.5 The influence of die material on the friction coefficient in the drawing of ETP Cu.From G. Baker and R. N. Wright, *Wire Journal International*, 25(6) (1992) 67.

marked increase in friction was observed for fine polycrystalline nib structure. The carbide dies displayed generally higher friction, although the friction was reduced for the case of a lower die angle, a phenomenon discussed in Section 5.4.

9.3. OTHER ELEMENTS OF DIE DESIGN

Beyond the basic aspects of die angle, reduction, and Δ, the following design features can be important.

9.3.1 Die entry geometry

Many die specifications involve loose and rather arbitrary guidelines for the bell portion of the die, yet there are some rather obvious factors to consider. First, a rapidly sloping, highly flared bell should minimize wire abrasion at the die entry in the event of die misalignment. This is particularly important if the die is intentionally misaligned as a means of controlling or imparting cast in the as-drawn wire. Beyond this, it has been argued that a relatively open bell facilitates lubricant entry that depends on the size of the entry area as opposed to the pressurization of the entering lubricant.

On the other hand, a lengthy, slowly tapering bell should produce a lubricant pressure buildup and foster hydrodynamic lubrication, when it can be achieved, similar to the effect of small approach angles. The same basic principles lie behind the tubular entry geometry advocated by Christopherson et al.[52] and in the early work of Wistreich.[53] Unfortunately, the bell geometry that assists lubrication is incompatible with the bell geometry that alleviates alignment complications.

9.3.2 Bearing length or land

One of the basic functions of the bearing length is to preserve the size of the die in the face of wear, and/or to allow refinishing of worn dies without necessarily increasing size. The geometrical practicality of this is readily seen in Figure 5.5. Bearing lengths in dies may vary from zero to as much as 200% of the wire diameter, although specifications of 30to 50% of the wire diameter seem to be the most common.

A principal objection to a large bearing length has been that it presumably adds to the frictional work in drawing the wire through the die, and hence adds to the draw stress. However, physical measurements do not show large drawing stress increases associated with the addition of moderate bearing lengths, presumably because wire contact with the bearing is minimal, or because there is no contact at all, as suggested by Figure 5.8. Certainly the notion of full die contact in the bearing seems unrealistic. More serious is the scraping of the wire on the bearing exit that undoubtedly occurs with die misalignment; however, this will be a potential problem regardless of the bearing length.

In short, the bearing length seems to be a relatively harmless expedient for obtaining extended die usage. Beyond this, the extended nib bulk required by the bearing and the back relief (see Section 9.3.3) strengthens the die vis-à-vis structural failure.

9.3.3 Back relief

Most die specifications are quite explicit about back relief and geometry, and the role of the back relief region in adding strength to the die has just been noted. Nonetheless specifications vary widely, and there is skepticism about the criticality of aspects of back relief. It has been supposed that back relief allows for some sort of gradual release of elastic energy at the die exit. Limited die–wire contact in the bearing makes this argument questionable, at best. On the other hand, a blended back relief should minimize scraping of the wire at the die exit, and the resultant surface abrasion and cast.

It can be strongly argued that the back relief should be specified with a radius as opposed to an angle. While this might complicate die manufacture, most of the alleged functions of the back relief would seem to be better performed with a curved profile.

9.3.4 Blending

The creation of an appropriate blend, or radius, at the transition from the approach angle to the bearing length is widely held to be a critical feature of die design. A number of general guidelines are set forth in the *Ferrous Wire Handbook*.[54] Relatively sharp blends are recommended for the drawing of high carbon steel and stainless steel, and substantial blending is cited for low carbon steel, copper, and aluminum. Too great a blend is felt to present the risk of unwanted cast. On the other hand, sharp blends are alleged to cause metal "pile-up," scratching, and out-of-roundness in softer metals, and well-blended contours are recommended. In any case, die wear will surely establish a form of blend between the approach angle and the bearing length, regardless of initial geometry.

Little theoretical work on die blends seems to have been done; however, the implications vis-à-vis Δ analysis are quite clear. A blend likely exposes the last stages of the drawing pass to the conditions of a lower die angle, and thus a lower Δ. This should reduce redundant work, die pressure, and centerline tension at the end of the pass. This can be a big factor in the case of light passes, where the nominal, overall Δ may be uncomfortably large. A significant portion of the pass may involve a lower range of die angles than the nominal approach angle specified for the drawing cone, especially if one remembers that the wire should first contact the drawing cone halfway down its length.

9.3.5 Sizing

The problem of appropriate sizing has been discussed by Ford and Wistreich as an aspect of die design.[55] These authors noted that die life is lost by sizing dies above the minimum allowable diameter, but precision sizing of dies close to the lower size limit may require added care and expense in die manufacture. The thinning behavior illustrated in Figure 5.8 may complicate the assessment of the minimum die size. A careful review of sizing practices is recommended to anyone concerned with improving die life, in a context where oversize as-drawn wire is a criterion for die replacement.

9.4. PRESSURE DIES

It has been noted that narrowly tapering bell geometry, low die angles, and, especially, Christopherson entry tubes may facilitate pressure buildup in the lubricant at die entry. Such pressure then should increase the quantity of lubricant passing between the die and the wire, and thus improve lubrication. These conditions lead to useful levels of residual lubricant on the wire.

Actually, the lubricant must have viscosity and wetting characteristics to facilitate such pressure buildup, and elevated drawing velocities may be necessary. Under ideal circumstances, hydrodynamic layers are expected to develop, as projected in Figure 7.1 and discussed in Section 8.1.4, with lubricant "dripping" from the wire surface upon exit from the die. Many drawing contexts offer the prospect of improved lubrication with increased lubricant pressure at die entry.

Pressure dies offer an efficient way to greatly enhance lubricant pressure at the die entry without the requirement of nearly ideal lubrication or high drawing speed. Figure 9.6 is a useful illustration of a pressure die system.[53] Two dies are usually involved. The first die, or the clearance die, has a

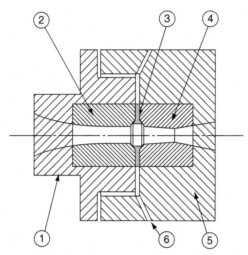

Figure 9.6 Pressure die system, where 1 is the rear casing, 2 is the clearance die, 3 is pressurized lubricant, 4 is the drawing die, 5 is the forward casing, and 6 is the relief vent. From J. G. Wistreich, *Wire and Wire Products*, 34 (1959) 1486, 1550.

diameter that may be a few percent larger than that of the wire or rod. The actual drawing occurs in the second die, which is called the drawing die. A chamber exists between the two dies. Lubricant is carried into this chamber by the wire or rod as it goes through the clearance die. The chamber, thus, becomes a reservoir for lubricant, with the lubricant becoming substantially pressurized. The pressurized lubricant is then able to pass through the drawing die. A pressure relief vent is generally provided to keep the lubricant pressure from getting too high and choking or even fracturing the system. Many pressure die designs have been developed, but the essentials are represented in Figure 9.6.

 9.5. DIE WEAR AND DIE LIFE

9.5.1 General wear analysis

Wire die wear is a special case of broader wear behavior involving sliding surfaces and the loss of surface material from adhesion and abrasion, among other factors. Equation 9.1, often called the Archard equation, is widely used for general analysis of wear behavior:

$$V_{\text{wear}}/L_{\text{sliding}} = q\ F/H, \qquad (9.1)$$

where V_{wear} is the volume of material worn away, $L_{sliding}$ is the distance of sliding, F is the force between the two sliding media, H is the hardness of the material being worn, and q is a proportionality constant.

With some geometrical simplification regarding the wear distribution, Equation 9.1 may be easily adapted to the case of wire die wear. The quantity $L_{sliding}$ is the length of wire drawn. The quantity V_{wear} is the area of the cylinder of contact between the wire and the die multiplied by the average depth of the die surface that has been worn away. This depth is $\delta/2$, with δ as the average die diameter *increase* due to wear. Finally, it is useful to replace the contact force with the product of the average die pressure, P, times the contact area. With these substitutions, Equation 9.1 can be restated in the following useful forms.

The increase in wire diameter due to wear may be expressed as:

$$\delta = P\,(2\,L_{sliding}\,q/H). \tag{9.2}$$

The length of wire that may be drawn for a given average die diameter increase is

$$L_{sliding} = (H\,\delta)/(2\,q\,P). \tag{9.3}$$

The die life in time, t_{life}, is $L_{sliding}/v$, where v is drawing speed, and hence:

$$t_{life} = (H\,\delta)/(2\,v\,q\,P). \tag{9.4}$$

Finally, the mass, M, represented by length, $L_{sliding}$, is

$$M = (\pi/4)\,L_{sliding}\rho\,d^2, \tag{9.5}$$

where ρ is wire density and d is as–drawn wire diameter.

For a specific drawing context, with H and q assumed constant and with δ as a given limit, Equation 9.4 can be simplified to:

$$t_{life} = Q_d/(vP), \tag{9.6}$$

where Q_d is a "constant," $(H\,\delta)/(2q)$. Therefore, the die life is seen to decrease as pressure increases and drawing speed increases.

9.5.2 Record keeping

Die life measurement and related data assessment require extensive record keeping as well as careful consideration of the criteria for die replacement. Die life should generally be recorded in terms of the length of wire drawn through the die, and length, mass, and time are generally relatable in Equations 9.2 to 9.6.

9.5.3 Criteria for die replacement

Many drawing shops simply replace dies when an as-drawn wire becomes oversized as per the value of δ in Equation 9.2. However, the development of the wear ring, as illustrated in Figure 5.5 and discussed in Section 5.5, generally precedes wear that enlarges the exit diameter. Thus, conservative drawing practitioners often use a degree of apparent wear ring development as the basis for taking a die out of the drawing line.

9.5.4 Die refinishing

In most instances, dies are returned to use after recutting and/or refinishing, typically to a larger size. Cost analysis of die usage generally involves the assumption of a certain number of recuts. The extent of recutting is related to the degree of wear, and dies taken out of service after extensive wear provide more limited recuts, and so on. With diamond nibs, dies taken out of service may be found to contain cracks. Depending on the location and extent of cracking, recutting options may or may not be limited.

9.5.5 Practical factors

Poor lubrication nearly always leads to shortened die life. Die wear will generally be increased by increased die pressure, as caused by lighter reductions, higher die angles, higher Δ values, and higher wire flow stress. Die wear may involve chemical reaction of the wire metal or the lubricant with the die. Die wear or fracture may reflect excessive drawing temperature related to lubricant breakdown or thermal stress in the die. Finally, die fracture may be due to inadequate compressive stress imposed by the die case on the nib.

9.6. PASS SCHEDULE CONCEPTS

9.6.1 Constant reduction

The simplest pass schedule concept is constant area reduction, r. This is especially straightforward when a gage system can be followed that involves the same area reduction. One such example is the Brown & Sharpe (B&S) or American Wire Gage (AWG) system, wherein each increase in gage number represents an area reduction of 20.70%, or an r value of 0.2070.

If the die angle remains the same, the reduction sequence involves a constant Δ value. For example, with a 12° included die angle, the B&S/AWG system involves a Δ value of 1.8 for each pass.

One basic advantage of this approach is that the wire emerging from each pass is a standard size to meet some standard specification. "Half-sizes" are often cited with die sizes to match. A half-size reduction in the B&S/AWG system involves an area reduction of 10.95%, or an r value of 0.1095.

To divide up a sequence of passes from A_0 to A_1 into n equal area reductions, true strain must be calculated, $\varepsilon_t = \ln(A_0/A_1)$, where strains may be directly added or subtracted. The overall true strain, in going from A_0 to A_1, is $\ln(A_0/A_1)$, and the true strain, ε_{tn}, in each of n passes is $[\ln(A_0/A_1)/n]$. The corresponding reduction, r_n, is obtained from the relation $r_n = 1 - \exp(-\varepsilon_{tn})$.

9.6.2 Constant Δ

While maintenance of constant reduction and die angle ensures a constant Δ value, it may be desirable to maintain a constant Δ value in the face of variable reductions. In such a case, Equation 4.9 provides a straightforward means for adjustments in die angle to achieve a constant Δ value in the face of varying reduction. For such purposes, Equation 4.9 can be rewritten as:

$$\alpha = r[1 + (1 - r)^{1/2}]^{-2}\Delta. \qquad (9.7)$$

One could, in principle, adjust the reduction to accommodate changes in die angle en route to constant Δ. However, there are few practical incentives for this adjustment.

The principle motivation for maintaining constant Δ is to maintain reasonably consistent drawing mechanics, such as consistent values of die pressure, redundant work, and centerline tension.

9.6.3 Constant ratio of draw stress to yielding/breaking stress

Pass schedules designed for maximum reduction in each pass should involve a constant ratio of draw stress to yielding/breaking stress, since that ratio most directly represents the risk of breakage in ostensibly sound rod or wire. It is generally advisable that this ratio be below 0.7, and a value of 0.6 probably represents a practical maximum for an aggressive reduction schedule.

The ratio of draw stress to yielding/breaking stress reflects the work hardening that may take place in going through the die. This complicates the draw stress analysis, since the work done during drawing reflects the average flow stress, not the yielding/breaking stress at the die exit. In any case, start with a rearrangement of Equation 5.13 where the draw stress, σ_d, can be expressed as a function of the average flow stress, σ_a:

$$\sigma_d = \sigma_a[(3.2/\Delta) + 0.9](\alpha + \mu). \tag{5.13}$$

In evaluating this effect, one can consider that a large amount of room temperature work-hardening data have been fitted with the simple expression $\sigma_o = k\varepsilon^N$, with σ_0 representing strength or true flow stress, ε as true strain, k as a strength coefficient, and N as the work-hardening exponent. By integration with $\sigma_o = k\varepsilon^N$, it can be determined that $\sigma_a = k\varepsilon^N/(N + 1)$. On this basis,

$$\sigma_d/\sigma_{01} = (N + 1)^{-1}[(3.2/\Delta) + 0.9](\alpha + \mu), \tag{9.8}$$

where σ_{01} is the flow stress or yield stress at the die exit. Therefore, given a practical maximum ratio, σ_d/σ_{01}, of draw stress to yield stress at the die exit, given a wire material property value of N and given values or α and μ, one can calculate the appropriate values of Δ and r. For example, if σ_d/σ_{01} is set at 0.6 and if N is 0.3, then with a 6° semi-angle and a coefficient of friction of 0.1, a Δ value of 1.1 and an r value of 0.32 or 32% are obtained.

It is tempting to use this value of r as a constant reduction through several passes; however, the value of N for the wire entering the die will generally decline with subsequent passes. If the above example calculation for N = 0.15 is repeated, a Δ value of 1.3 and an r value of 0.275 or 27.5% are obtained.

9.6.4 Constant ratio of draw stress to average flow stress

Since the database and calculations required for a constant draw stress to yielding/breaking stress ratio are somewhat tedious, a useful alternative may be to simply use Equation 5.13 cited above. This approach ignores the contribution of work hardening to the maximum practical reduction, but it provides a more conservative basis to design a pass schedule for maximum reductions. For example, if σ_d/σ_a is set at 0.6, then for the case of a 6° semi-angle and a coefficient of friction of 0.1, a Δ value of 1.6 and an r value of 0.235 or 23.5% are obtained with Equation 5.13. It could be argued that projections for allowable reductions that ignore work hardening are too conservative; however, that can be recognized when making and using the calculation.

9.6.5 Constant draw stress and constant heating

Practical control of temperature in drawing is more easily achieved if the amount of heat generated in each drawing pass is controlled. As shown in Equation 6.1, for adiabatic conditions the temperature increase is proportional to the drawing stress. Therefore, maintenance of the same draw stress for each pass should ensure consistent heating for each pass. If that heating can be substantially removed between passes, then a thermally stable drawing line can be maintained.

As drawing ensues, work hardening causes the flow stress and the draw stress to increase proportionately. Therefore, to maintain a constant drawing stress in the face of this flow stress increase, the reduction per pass must be reduced. A pass schedule involving a steady decrease in per-pass reduction to offset work hardening is called a "tapered" pass schedule.

Since the reduction impacts redundant work, it is expedient to consider modifying Δ with increasing flow stress to achieve consistent draw stress and heating per pass. If the die angle and the coefficient of friction are assumed to remain stable, then Equation 5.13 indicates that constant draw stress will be equivalent to constancy of the expression $\sigma_a [(3.2/\Delta) + 0.9]$, which in turn equals $\sigma_d/(\alpha + \mu)$.

If it is assumed that the flow stress of the wire corresponds to the relationship $\sigma_o = k\varepsilon^N$, then the average flow stress in a given pass, n, should correspond to the relationship:

$$\sigma_{an} = k[\ln(A_0/A_n)]^N, \tag{9.9}$$

where σ_{an} is the average flow stress in a given pass, n, where A_0 is the cross-sectional area of the wire or rod at the beginning of the pass schedule, and where A_n is the cross-sectional area after pass n. On this basis, Equation 5.13 can be modified to the form

$$\sigma_d/[k(\alpha+\mu)] = \text{Constant} = [\ln (A_0/A_n)]^N\{[(0.8/\tan\alpha) \ln (A_{n-1}/A_n)]+0.9\} \tag{9.10}$$

Equation 9.10 allows for the solution of a series of cross-sectional areas associated with the desired tapered pass schedule; that is, one selects the desired value of σ_d (or temperature increase), calculates the value of the Constant, and solves for A_n, as a function of N and A_{n-1}. For the first pass, A_{n-1} is A_0, the starting cross section, and A_1 is a function of N. For the second pass, A_{n-1} is A_1, and so on. A detailed calculation like this is presented below in Section 9.8.

9.6.6 The relationship of pass schedules to slip

"Slip" is the relative motion of the wire and the capstan surface as related to differentials between the drawing speed and the faster capstan surface speed; that is, the exit velocity of the wire, V_1, and the surface velocity of the pulling capstan, $V_{capstan}$ do not have to be the same. Their difference, or $V_{capstan} - V_1$, is the amount of slip, or

$$\% \, slip = 100(V_{capstan} - V_1)/V_{capstan}. \tag{9.11}$$

The exit velocity of the wire for sequential passes follows the simple relation

$$V_{n+1} = V_n/(1-r), \tag{9.12}$$

and

$$V_{capstan} = \pi \, D \, \omega, \tag{9.13}$$

where D is capstan diameter and ω is capstan revolutions per unit time.

It is possible to set or adjust capstan surface velocity by fixing D, as with a step-cone machine (see Figure 3.2), or by adjusting ω. It is possible to have $V_{capstan}$ equal V_1, in which case we would have "no slip" drawing practice.

Designing in some slip allows the easy accommodation of changes in wire velocity due to die size variations and die wear. Some degree of capstan wear is also accommodated. On the other hand, slip promotes capstan wear and compromises wire surface quality.

Engineered slip schedules can range from a small percent per pass to, say, 20 percent per pass. Small degrees of slip increase the risk of drawing breaks, and larger amounts of slip increase wear and surface damage. The engineering of slip schedules is an aspect of pass design. In some cases, the accumulated slip over several passes is factored into pass design.

Slip interacts with drawing stress, and it is useful to know how the capstan exerts a pulling stress on the wire in the face of slip and related friction. Classical belt drive formulas are applicable. A relationship of drawing stress, σ_d, to the stress coming off the capstan as back stress at the next die, σ_b, is illustrated in Figure 9.7. The stress ratio is

$$\sigma_d/\sigma_b = \exp(\varphi\mu) = \exp(2\pi N\mu), \tag{9.14}$$

where φ is the angle of wire contact, μ is the coefficient of friction between the wire and the capstan, and N is the number of wraps of the wire around the capstan.

With an adequate number of wraps, σ_b can be acceptably small. For example, if N is three and μ is 0.1, the back stress going into the next die

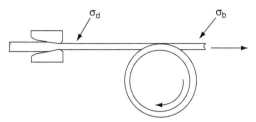

Figure 9.7 Stresses going onto and coming off of a capstan.

will be only 15% of the previous drawing stress, σ_d. Such values can be significant; however, and may be incorporated into draw stress calculations by using Equation 5.20.

9.7. DRAWING PRACTICE AND CAST AND PITCH CONTROL

9.7.1 Definitions and origins

Wire often assumes a helical configuration, going onto spools or into containers, and may exhibit a helical configuration coming off of spools or out of containers. Assuming the helical configuration has a longitudinal axis, the cast is described by the radius of curvature about that axis, and the pitch is the advance of the wire along that axis in one revolution.

In many cases, cast and pitch are undesirable and are eliminated or at least substantially reduced by straighteners, although straightening may produce residual stresses that lead to subsequent stress relaxation and creep, re-establishing a degree of out-of-straightness. Such residual stresses, combined with the stresses produced by coiling, may cause stress relaxation and creep, leading to cast and pitch upon uncoiling.

On the other hand, welding wire and other payoff-sensitive wire products are often produced with a desired cast and pitch to allow the wire optimal entry into a welding or fabrication station. Such cast and pitch can be imposed by passing the wire through tooling that imposes the helical configuration prior to spooling or containerization. In some contexts, cast and pitch control has been attempted by alignment or controlled misalignment of the finishing die.

9.7.2 Drawing issues

The mechanics of cast and pitch development or control on a wire with a truly round cross section, a homogenous microstructure and state of work hardening, and no residual stress would seem relatively straightforward.

Little should be required beyond dealing with wire from a "sprung back" configuration (see Sections 18.2 and 18.3) subsequent to bending and twisting. However, drawn wire is generally not so geometrically and mechanical neutral, and the following considerations are pertinent.

9.7.2.1 The state of residual stress

The state of residual stress expected at the wire surface (see Section 6.4.3) will lead to easy yielding in the near-surface regions along the outer bend radius, and a resistance to yielding in the near-surface regions along the inner bend radius. The neutral axis will shift from the centerline. This can be dealt with if the residual stress pattern is uniform longitudinally.

9.7.2.2 Wire with a seam

In wire with a seam (i.e., with a core of flux or alloy powder), the seam must affect cast. In general the seam will result in easier bending and increased cast. Any circumferential variation in drawing response due to the seam will also be significant.

9.7.2.3 Sources of circumferential variation in drawn wire

The variations to be considered are those of redundant work and residual stress caused by the following:
- High Δ drawing, leading to greater degrees of redundant work and a greater potential for non-uniformity
- Worn dies, since wear and the related interference with lubrication may not be circumferentially uniform
- Die misalignment, including intentional misalignments to create cast
- Misalignments of the "wire route" with the die holder (influences of capstans, guides, etc.)
- Dies drilled off center, or with asymmetric blends
- Vibrations along the wire route

9.7.2.4 Longitudinal variations

Drawing temperature fluctuation is a source of longitudinal variation that can affect cast and pitch control with lubrication variation and flow stress variation.

9.8. QUESTIONS AND PROBLEMS

9.8.1 Much can be learned from drawing a realistic die design, particularly since many die illustrations or iconic representations are exaggerated. Make a drawing (say, 40×full scale) of a 12° included angle die with a 30% bearing intended for drawing with a 20% reduction to 1 mm diameter. Be quantitative with the drawing cone and the bearing; do not put in a blend between the drawing cone and the bearing and make certain that wire–die contact will first occur half-way through the drawing cone. What is the contact length for the wire?

Answer: In making the drawing, it may be noted that the initial diameter can be calculated to be 1.118 mm for a 20% reduction. From basic trigonometry, the contact length will simply be $[(d_0 - d_1)/2]/\sin 6°]$, which is 0.564 mm.

9.8.2 On average, 2000 kg of a certain steel can be drawn through a 1 mm die before it is necessary to change the die. What is the length of wire that can be drawn through the die? If wear rates remain about the same, estimate the mass of steel that can be drawn through a 0.5 mm die. If the drawing speed is 600 m/min, what will be the lifetime of the die in hours?

Answer: The mass equals the density (see Section 19.2) times the volume. The volume is the cross-sectional area times the length. On this basis, the length can be solved at 3.24×10^5 m. This length is a basic measure of wear, and will not (in principle) change with diameter. Therefore, the mass to be drawn through a 0.5 mm die will be only one-quarter that for the 1.0 mm die, or 500 kg. The time consumed in drawing 3.24×10^5 m of wire is length divided by drawing speed, leading to 540 min, or 9 hours.

9.8.3 A certain light-reduction finishing pass involves higher die wear than heavier, earlier passes. Explain.

Answer: A light reduction may be associated with a high Δ value. Equation 5.18 indicates that an increased Δ value will increase die pressure, which in turn should increase die wear.

9.8.4 Equation 9.10 provides a basis for calculating the sequence of reductions in a tapered drawing schedule. Consider a certain steel with a work-hardening exponent, N, of 0.2. If the included die angle is 12°, and if the constant in Equation 9.10 is 1.5, what will be the first reduction of the tapered drawing schedule?

Answer: For the first pass, A_n is A_1 and A_{n-1} is A_0. The equations can be simplified by substituting "x" for $\ln(A_0/A_1)$. With iteration, x can be solved for as 0.166, so that A_0/A_1 is 1.18 and the reduction is 0.153, or 15.3%.

9.8.5 A condition of 5% slip exists on a 0.3 m diameter capstan rotating 600 revolutions per minute. What is the speed, in m/min, for the wire as it exits the previous die?

Answer: Using Equations 9.13 and 9.11, V_c can be calculated to be 565.5 m/s and V_1 to be 537.2 m/s.

9.8.6 A certain capstan maintains a coefficient of friction of 0.10 on a wire that is wrapped three times. If the drawing stress on the previous die is 700 MPa, what will be the back stress on the next die? What would be the value of the back stress if there were only two wraps?

Answer: Using Equation 9.14, the back stress for three wraps is 106 MPa, and for two wraps it is 199 MPa.

CHAPTER TEN

Shaped Dies and Roller Dies

Contents

10.1. DRAWING SHAPES WITH ONE-PIECE DIES

10.1.1 Introduction

The drawing of shaped bar, rod, and wire stock with one-piece dies is an important near-net-shape technology, offering sizing, dimensional control, and surface quality not generally achievable with rolling and extrusion. The shaped rod may be directly useful in electric power transmission configurations, or it may be sectioned as a preform for items such as nuts, firearm parts, and so on.

The mechanics of shape drawing are much more complex than for axisymmetric and plane strain drawing, and industrial practice has been based substantially on empirical observations. A brief summary of die and pass schedule design for shape drawing has been published by Perlin and Ermanok.[56,57] In this chapter, we will focus on those guidelines and review some important engineering observations from the work of Wright, Martine, and Yi.[58-60]

10.1.2 General guidelines

It can be said with certainty that better results will be achieved if the incoming cross-section geometry, or the preform is optimized; that is, better results can be achieved if the geometrical transition in a given pass is limited. A shaped die drawing sequence with a rolled or extruded product with generically similar geometry to the end product should be initiated, if at all possible.

Next, it is important that the incoming rod contact the drawing channel in a planar locus; that is, all points on the periphery of the incoming rod should contact the die at the same time. Beyond this, many shaped products have corners, and these corners should be sharpened gradually, from pass to pass, assuming sharp corners are desired. Lower Δ values, or larger reductions and smaller die angles, ease drawing conditions by lowering die pressure and reducing redundant work. However, as shown in this chapter, higher Δ values may be necessary to achieve sharp corners.

It is important to minimize, or at least be aware of, non-uniform or extraneous plastic deformation. In addition to redundant work associated with higher Δ drawing passes and more aggressive shape transitions, circumferential Δ variations due to non-planar entry and die/workpiece misalignment should be avoided. Finally, twisting may be a problem with some pass-to-pass geometry transformations.

10.1.3 Die and pass design

As previously stated, it is important that the entry locus of the rod into the die be planar as well as the exit locus. Thus it is desirable to have a linear, longitudinal transition between the entry shape and the exit shape, not only for a single pass, but for a sequence of passes. Figure 10.1 illustrates the design of a profile sequence from an initial rectangular cross section to a final "T"-shaped cross section over thirteen passes.[56,57] Such transitions involve a range of die angles within each pass.

Figure 10.2 illustrates the range of die angles involved in two different dies with simple one-pass transitions from a round entry geometry to an hexagonal exit shape with an "across-the-flats" diameter of 10.31 mm.[58-60] One of the dies involves lower die angles, and the other involves higher die angles. (Data involving these die designs are discussed in Section 10.1.5.) The end-on section shows the transition from either of two circular entry

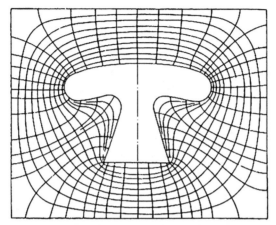

Figure 10.1 Design of a shaped wire profile sequence from an initial rectangular cross section to a final "T-shaped" cross section, over thirteen passes. From J. L. Perlin and M. Z. Ermanok, Theory of Wire Drawing, *Metallurgia*, USSR, (1971), and T. A. Kircher and R. N. Wright, Wire Journal International, 16(12) (1983) 40.

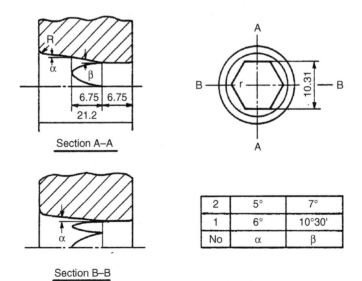

No	α	β
1	6°	10°30'
2	5°	7°

Figure 10.2 Designs for two dies providing transitions from round entry geometry to hexagonal shape. From P. B. Martine, An Analysis of the Drawing of Hexagonal Bar from Round Stock, M. S. Thesis, Rensselaer Polytechnic Institute, 1982

cross sections to a final hexagonal cross section. Section B-B, an "across-the-corners" section, reveals a die semi-angle, α, which is the angle between the corner locus and the centerline. Section A-A, an "across-the-flats" section, reveals a semi-angle, β, which is the angle between the locus at the center of the flat sides and the centerline. Two sets of α and β values are tabulated in Figure 10.2 for die designs 1 and 2. The β values are higher than the α values.

The reduction for the passes is calculated in the usual manner, and an average Δ or Δ_{ave} can be calculated from the average of the two die semi-angles. For shape drawing, it is a reasonable expedient to use the Δ_{ave} values in place of Δ in the various formulas for draw stress, redundant work, average die pressure, centerline tension, and so on.

10.1.4 Redundant work patterns

In general, redundant work is higher in the outer, nearer-surface regions, just as in round-to-round drawing. However, it is especially high at corner regions. Figure 10.3 displays a qualitative image of this type of hardness distribution, with the hardness plotted vertically upon the hexagonal cross section.[58-60]

10.1.5 Laboratory observations

Wright, Martine, and Yi studied the drawing mechanics for single round-to-hexagonal drawing passes on free-machining brass using die designs 1 and 2 shown in Figure 10.2[58-60]. Drawing forces were measured and cross-sectional

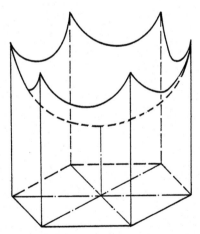

Figure 10.3 Qualitative image of the hardness distribution (plotted vertically) in a cross section of a cold drawn hexagonal rod. From Y. Yi, Metal Flow in the Drawing of Shaped Bar, M. S. Thesis, Rensselaer Polytechnic Institute, 1981.

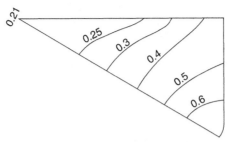

Figure 10.4 Lines of constant strain in 30° segment of a cross section of cold drawn hexagonal rod. From Y. Yi, Metal Flow in the Drawing of Shaped Bar, M. S. Thesis, Rensselaer Polytechnic Institute, 1981.

strain patterns were determined by comparing micro-hardness values with a reference curve, displaying basic hardness versus true strain relations for the brass workpiece material. Figure 10.4 shows lines of constant strain in a basic 30° section of the hexagonal cross section. From this data it is possible to compute the redundant work factor for the pass. Lubrication was achieved with a high viscosity oil. Representative data from this study are illustrated in Table 10.1.

It is clear that the higher die angle and higher Δ pass involved much more redundant work. The increased redundant work undoubtedly involves increased die pressure, and the increased redundant work and die pressure appeared to be associated with better corner filling (sharper as-drawn corners, as indicated by the larger value of D_c for the higher Δ pass).

This study noted a brighter surface in the middle of the flats than near the corners. In this context friction coefficients were measured for round-to-round drawing, round-to-hexagon drawing, and plane strain strip drawing, using the same high viscosity oil. The coefficient of friction in round-to-round drawing was only one-third the value of the coefficient of friction for round-to-hexagonal drawing. On the other hand, the coefficient of friction for plane strain strip drawing was over twice that of round-to-hexagonal

Table 10.1 Data for drawing 10.31 mm (across-the flats) hexagonal rods from round rods

Die Number	$\ln(A_0/A_1)$	D_f (mm)	D_c (mm)	σ_d/σ_a	Φ	Δ_{ave}	μ
1	0.196	10.312	11.56	0.45	1.31	2.94	0.11
2	0.196	10.325	11.43	0.45	1.08	2.14	0.12

Note: The across-the-flat and across-the corner diameters are designated by D_f and D_c, respectively. Data from the work of Wright, Martine, and Yi[58–60].

drawing. In plane strain strip drawing it appeared that the lubricant escaped out the unconstrained sides of the die–strip interfacial region, leading to a very high coefficient of friction. In hexagonal drawing, it appears that corners of the hexagonal die act as reservoirs for lubricant squeezed out from the flat sections. In round drawing, no such lubricant escape occurs, leading to a lower coefficient of friction. Finally, the brighter surface of the middle of the flats may be attributed to the thin film lubrication that develops in the middle of the flats when the lubricant is pushed to the corners.

10.2. DRAWING WITH UNPOWERED ROLLER DIE SYSTEMS

10.2.1 Unpowered roll configurations

Some simple roll configurations are shown in Figure 10.5a, b, and c. Figure 10.5a shows a basic open rolling, or roll flattening configuration. Figure 10.5b involves the presence of edging rolls. Four roll configurations such as those of Figures 10.5b and 10.5c are often referred to as "turksheads." Much more

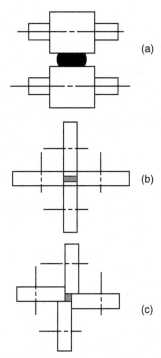

Figure 10.5 Simple roller die configurations: a) basic open rolling or roll flattening, b) four-roll configuration with edging rolls, c) four-roll "turkshead" configuration.

complex configurations have been developed, involving multiple passes. Analysis of these types of configurations is beyond the scope of this book, as was analysis of complex one-piece die sequences en route to complex shapes. We will, however, address the rolling of rectangular cross sections in single passes. This is a major technology, and sequences of such passes can be employed to many ends.

10.2.2 Comparison to drawing with solid dies

Chapter 5 presented several analyses based on the concepts of uniform work, redundant work, and friction work, and the contributions of all three to the drawing stress, which in turn is equal to the total work per unit volume. Moreover, these analyses were generally simplified by use of the deformation zone geometry index Δ. This approach will continue to be used.

Drawing or pulling the workpiece through rollers to effect a change in cross-sectional area or shape differs from simple drawing in that very little friction is created by the rollers, and a lubricant may not be necessary. Hence, it is reasonable to eliminate friction work from the analysis. Equation 5.5, expressing the role of uniform work, can now be reintroduced as

$$w_u = \sigma_a \ln[1/(1-r)]. \tag{10.1}$$

Equation 5.7, expressing the role of non-uniform work, can now be reintroduced as

$$w_r = \sigma_a (\Phi - 1) \ln[1/(1-r)], \tag{10.2}$$

and therefore

$$\sigma_d = \sigma_a \Phi \ln[1/(1-r)]. \tag{10.3}$$

It remains necessary to quantify Φ, the redundant work factor, using Δ.

10.2.3 Two-roll systems and Δ

The two-roll systems are a form of flat rolling, and the Δ value for *plane strain* flat rolling, Δ', is expressed as

$$\Delta' = [t_0/(4R_r r)]^{1/2}(2-r), \tag{10.4}$$

where t_0 is the initial thickness of the entering strip and R_r is the roll radius. A geometrical description of this form of Δ is shown in Figure 10.6.[61]

Plane strain occurs where there is no spreading or increase in rectangular rod width during the drawing of the rod through the rolls. This will be

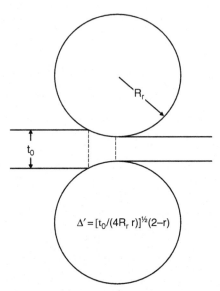

$$\Delta' = [t_0/(4R_r\, r)]^{1/2}(2-r)$$

Figure 10.6 Configuration for Δ' (deformation zone shape parameter) in rolling. From W. A. Backofen, *Deformation Processing*, Addison-Wesley Publishing Company, Reading, MA, 1972, 89. Copyright held by Pearson Education, Upper Saddle River, NJ, USA.

essentially the case when the rectangular rod or strip width is ten times or more its thickness. Plane strain assumptions are often used on an approximate basis when the strip width is three times or more its thickness. The ratio of workpiece width to thickness is called the *aspect ratio*.

To a first approximation, we can use Equation 5.8 to estimate Φ on the basis of Δ', so that

$$\Phi \approx 0.8 + \Delta'/(4.4). \tag{10.5}$$

Now, for the two-roll system with plane strain,

$$\sigma_d \approx \sigma'_a[0.8 + \Delta'/(4.4)]\ln[1/(1-r)], \tag{10.6}$$

where σ_a' is the average flow stress for plane strain deformation and is equal to $(1.15)\sigma_a$. Thus, in terms of σ_a, Equation 10.6 can be rewritten as

$$\sigma_d \approx \sigma_a[0.92 + \Delta'/(3.8)]\ln[1/(1-r)]. \tag{10.7}$$

For significant side spreading in a two-roll system, it is recommended that Equation 10.6 be modified to reflect σ_a and not σ_a'. Thus the drawing stress for a two-roll system with significant side spreading is estimated as

$$\sigma_d \approx \sigma_a[0.8 + \Delta'/(4.4)]\ln[1/(1-r)]. \qquad (10.8)$$

Equation 10.8 should be used when the rectangular rod width is the order of the thickness. A detailed treatment of side spread prediction in rolling is given in Chapter 17.

It can be argued that Equation 10.8 does not include frictional work due to spreading of the workpiece, along the roll axis, in the transverse direction of the workpiece. It seems unlikely that such an effect will increase σ_d more than a very few percent, however.

10.2.4 Four-roll systems and Δ

Roller die systems often involve additional rolls on the side. When such rolls achieve little actual deformation in the transverse direction, there is no reason not to use Equations 10.6 and 10.7. Such an assumption seems reasonable for aspect ratios of three or more.

When the side rolls do achieve transverse reduction, it is recommended that one use an average value of Δ', where Δ'_{ave} is the numerical average of Equation 10.4 applied in both transverse directions; that is, consider a roller die system with mutually orthogonal axes x, y, and z, let the rolling direction be z, the "main" roll axis be x, and the side roll axis be y. On this basis,

$$\Delta'_{ave} = 1/2\left\{[w_{0x}/(4R_s r_{xz})]^{1/2}(2-r_{xz}) + [t_{0y}/(4R_r r_{yz})]^{1/2}(2-r_{yz})\right\}, \quad (10.9)$$

where w_{0x} is the entering strip width (in the x direction), t_{0y} is the entering strip thickness (in the y direction), R_r is the "main" roll radius, and R_s is the side roll radius. The values for r_{xz} and r_{yz} are $r_{xz} = (w_{0x} - w_{1x})/w_{0x}$ and $r_{yz} = (t_{0y} - t_{1y})/t_{0y}$, where w_{1x} is the exiting strip width and t_{1y} is the exiting strip thickness.

The drawing stress for this case where side rolls do achieve reduction would be

$$\sigma_d \approx \sigma_a[0.8 + \Delta'_{ave}/(4.4)]\ln[1/(1-r)]. \qquad (10.10)$$

10.3. "DRAWING" WITH POWERED ROLLER DIE SYSTEMS

10.3.1 Comparison to rolling

The word "drawing" in the title of this section is put in quotation marks because this technology is really not drawing at all, since there is no pulling activity (front and back tension in rolling notwithstanding). It is important to consider it at this point, however, since it is a clear alternate technology to that of pulling through rollers. We will restrict ourselves, however, to systems where only the two "main" rolls are powered.

10.3.2 Roll force analysis for plane strain rolling

Basic analysis of powered, plane strain rolling, on any scale, involves analysis of forces on the rolls and associated calculations of torque and power requirement. There can be a strong roll of friction in the analysis, unlike the technologies discussed in Section 10.2. The simplest roll force formula is

$$F_r = L_r w \sigma_a [1 + (\mu L_r)/(t_0 + t_1)], \qquad (10.11)$$

where F_r is the roll force, L_r is the length of contact of the rolls with the workpiece in the longitudinal direction, w is the workpiece width, and t_0 and t_1 are the initial and final strip thicknesses, respectively. The value of L_r is given by

$$L_r = [R_r(t_0 - t_1)]^{1/2}. \qquad (10.12)$$

The roll force can be used to calculate the torque and power involved with the rolling pass. The total torque, T_r, on two rolls, is given by

$$T_r = F_r L_r, \qquad (10.13)$$

and the power consumed, P_r, is given by

$$P_r = T_r \omega, \qquad (10.14)$$

where ω is the roll speed in revolutions per unit time. The actual power consumed for the rolling mill, overall, will be larger, consistent with the electromechanical efficiencies of the overall drive system.

The plane strain analysis is pertinent when the workpiece is constrained from spreading by edging rolls, or because of an aspect ratio of ten or more. And, as noted previously, plane strain assumptions are often used on an approximate basis when the aspect ratio is three or more.

10.3.3 Roll flattening of rod and wire

In this technology, round wire is generally rolled between flat rolls to achieve a flattened shape generically like that illustrated in Figure 10.5a. The aspect ratio is too low to use plane strain analysis, and the dominant issue is the prediction of and management of spread. A rather comprehensive treatment of practical spread analysis is presented in Chapter 17. For the purposes of this chapter, it may be noted that spread, S, may be defined as w_1/w_0, where w_1 is the width of the rod after the pass and w_0 is the width before the pass. For an approximate analysis, for a given roll diameter, state of lubrication, aspect ratio, temperature, roll speed, and so forth, a spread index, n_s, may be calculated where

$$n_s = [(w_1/w_0)/(A_1/A_0)]^{1/2}. \qquad (10.15)$$

Higher values of n_s are associated with increased spreading and less elongation, and lower values of n_s are associated with decreased spreading and increased elongation. A value of n_s of one indicates roughly equal tendencies toward spread and elongation. The trade-off between spread and elongation reflects constant volume of the workpiece. Sperduti published roll flattening data which, for 20 cm diameter rolls and starting rod diameters the order of 4 mm to 10 mm, imply n_s values of roughly 1.04.[62]

Modeling of the roll force for roll flattening is complicated by the spreading. However, if one ignores friction and redundant work, it is clear that the work per unit volume in a rolling pass will be $\sigma_a \ln(A_0/A_1)$. Since the volume rolled per unit time is about $\pi R_r \omega (A_0 + A_1)$, then the work per unit time, or power, consumed by the metal working, itself, is

$$P_r = \sigma_a \pi R_r \omega (A_0 + A_1) \ln(A_0/A_1), \qquad (10.16)$$

and based on Equation 10.14,

$$T_r = \sigma_a \pi R_r (A_0 + A_1) \ln(A_0/A_1). \qquad (10.17)$$

 ## 10.4. ROLL GAP ISSUES

In the simplest concept, it can be imagined that the roll gap is literally the workpiece thickness to be expected upon rolling. However, it is often the case that the rolls are pushed apart somewhat by the roll force, and the workpiece thickness is larger than the initial roll gap setting. This situation is described in Figure 10.7, showing the *plastic curve* and the *elastic curve*, and their point of intersection, which describes the actual rolling conditions.[63]

The plastic curve is a plot of rolling force versus the change in workpiece thickness; that is, if the initial workpiece thickness is t_0 then the rolling force is a function of $(t_0 - t_1)$, where t_1 is the as-rolled thickness. The elastic curve is a plot of rolling force versus the increase in the roll gap, or $(t_1 - g)$, where

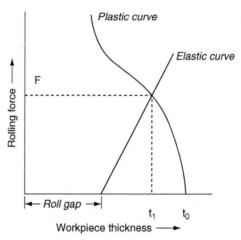

Figure 10.7 Plastic and elastic curves for rolling, intersecting at actual rolling force and as-rolled thickness. From G. E. Dieter *Mechanical Metallurgy*, Third Edition, McGraw-Hill, Boston, MA, 1986, 606. Copyright held by McGraw-Hill Education, New York, USA.

g is the initial roll gap distance. Under actual rolling conditions, both forces are the same, namely the force value at the point of intersection of the elastic and plastic curves. The workpiece thickness at that point is the as-rolled thickness.

The slope of the elastic curve reflects the elasticity of the mill structure. The plastic curve is difficult to model in a practical, effective way. However, the general construct of Figure 10.7 can be quantified empirically for a given mill and process and can be of great value in process setup.

10.5. QUESTIONS AND PROBLEMS

10.5.1 Figure 10.2 and Table 10.1 present information leading to values of Δ_{ave}. Using the die semi-angles presented in Figure 10.2, confirm the corresponding values of Δ_{ave}. Also, compare the values of Δ_{ave} to an optimum value of Δ_{ave} based on Equation 5.16.

Answer: The reductions are 0.178, as converted from the ln (A_0/A_1) expression. For die one, the average of the two die semi-angles is 8.25° and the value of Δ_{ave} is 2.95, nearly the same as in Table 10.1. For die two, the average of the two die semi-angles is 6° and the value of Δ_{ave} is 2.14, the same as in Table 10.1. Using Equation 5.16, the optimum value of Δ_{ave}, for draw stress minimization, is calculated to be 2.90, which is essentially the same for die one.

10.5.2 Equation 10.8 expresses the drawing stress for a condition of significant side spread in a two-roll system. Consider a workpiece 3 mm thick rolled to a reduction of 15%, with significant side spread, on a mill with a roll diameter of 10 cm. Calculate the ratio of the drawing stress to the average flow stress.

Answer: Taking care to convert the diameter to a radius, the value of Δ' may be calculated from Equation 10.4 as 0.585. The ratio of the drawing stress to the average flow stress can then be calculated to be 0.152.

10.5.3 Consider a four-roll system with a principal roll diameter of 4 cm and a side roll diameter of 2 cm. If the workpiece enters with dimensions of 2 cm × 3 mm, and exits with dimensions of 1.8 cm and 2.8 mm, what should be the ratio of the drawing stress to the average flow stress?

Answer: The value of Δ'_{ave} may be calculated from Equation 10.9 as 2.85. Using Equation 10.10, the ratio of the drawing stress to the average flow stress can then be calculated to be 0.252.

10.5.4 Consider Equation 10.16, and consider a workpiece with an initial cross-sectional-area of 2 cm^2 that is rolled to a 10% reduction on a mill with 5 cm diameter rolls, operating at 100 revolutions per minute. If the average flow stress of the workpiece is 700 MPa, how many watts will be consumed in this operation?

Answer: Putting all of the values into Equation 10.16, the power consumed may be calculated to be about 3660 Nm/s or about 3.565 kW. Various electromechanical inefficiencies will lead to additional power consumption.

CHAPTER ELEVEN

Mechanical Properties of Wire and Related Testing

Contents

Wire Technology
ISBN 978-0-12-382092-1, DOI: 10.1016/B978-0-12-382092-1.00011-7

11.1. THE FLOW STRESS OF THE WIRE

As noted in the previous chapters, many practical calculations for wire process analysis require a knowledge of the wire flow stress. The flow stress is simply a stress at which the wire yields, or deforms plastically. Yield criteria are equations that incorporate a known value of the flow stress or yield stress, and allow the calculation of states of stress that are necessary for yielding. There are two widely used yield criteria, the Tresca criterion and the von Mises criterion. While these criteria can involve complex descriptions of stress state, they can be used simply yet quite usefully in wire drawing analysis.

11.1.1 The Tresca criterion

In its simplest form, the Tresca criterion can be written as

$$\sigma_o = \sigma_I - \sigma_{III}, \tag{11.1}$$

where σ_o is the flow stress, σ_I is the most tensile or least compressive normal stress, and σ_{III} is the most compressive or least tensile normal stress. Stresses such as σ_I and σ_{III} are called principal stresses because they act on faces that have no shear stress acting upon them. There is actually an additional principal stress, σ_{II}, but only the extreme-valued normal stresses, σ_I and σ_{III}, are involved in the Tresca criterion.

The Tresca criterion is equivalent to saying that yielding will occur at a critical value of the maximum shear stress, consistent with micromechanical behavior of crystals, involving slip and dislocation motion.

Equation 11.1 indicates that when the difference between σ_I and σ_{III} reaches the value of the flow stress, σ_o, then yielding and plastic flow will occur. The values of σ_I and σ_{III} represent the applied stress state, and the flow stress, σ_o is a material property, directly obtained from the tensile test, as discussed in Section 11.1.4.

11.1.2 The von Mises criterion

The von Mises criterion involves all three principal stresses, and can be written as

$$\sigma_o = 2^{-1/2}\left[(\sigma_I-\sigma_{II})^2 + (\sigma_{II}-\sigma_{III})^2 + (\sigma_{III}-\sigma_I)^2\right]^{1/2}. \tag{11.2}$$

When the value of the right side of Equation 11.2 reaches the value of the flow stress, σ_o, then yielding and plastic flow will occur.

The von Mises criterion is equivalent to saying that yielding will occur at a critical value of the elastic energy in the material. Both the Tresca and the von Mises criteria provide useful engineering predictions of yielding, and analysts tend to use the criterion that is most convenient or most consistent with the overall mechanical context.

11.1.3 Effective stress and effective strain

When continuing plastic flow occurs, the stress combinations implied can be stated by the von Mises criterion, equating σ_o to an instantaneous or current value. This value is commonly called the flow stress and/or the effective stress, as shown in Equation 11.2.

In isotropic materials, or materials with the same yield strength in all directions, a quantity called the effective strain, or ε_o, can be coupled to the effective stress. The effective strain is defined by the relationship

$$\varepsilon_o = (2^{1/2}/3)\left[(\varepsilon_I-\varepsilon_{II})^2 + (\varepsilon_{II}-\varepsilon_{III})^2 + (\varepsilon_{III}-\varepsilon_I)^2\right]^{1/2}. \qquad (11.3)$$

The terms ε_I, ε_{II}, and ε_{III} are principal strains as well as normal strains acting in the same directions and on the same faces as the principal stresses.

11.1.4 The flow curve and the tensile test

A plot of effective stress versus effective strain is called the flow curve, and flow curve data can be used to address a wide variety of complex problems of plastic flow. It is most useful that, prior to necking, the true stress-strain curve produced by the tensile test is *literally* the effective stress versus effective strain flow curve.

This equivalency of the stress-strain curve and the effective stress versus effective strain flow curve is readily demonstrated. Prior to necking, the values of the principal stresses, σ_I, σ_{II}, and σ_{III} in a tensile test are, respectively σ_t, 0, and 0, where σ_t is simply the true stress along the tensile axis. If one inserts these values into Equation 11.2, it is shown that $\sigma_o = \sigma_t$. (Once necking occurs, the neck geometry causes radial tensile stress, perpendicular to the tensile axis, and σ_t exceeds σ_o somewhat.)

Similarly, the values of the principal strains, ε_I, ε_{II}, and ε_{III}, after yielding in a tensile test are, respectively ε_t, $-0.5\varepsilon_t$, and $-0.5\varepsilon_t$, where ε_t is simply the true strain along the tensile axis. If these values are inserted into Equation 11.3, it is shown that $\varepsilon_0 = \varepsilon_t$.

In most wire process analysis, it is simply necessary to know the flow stress associated with the current microstructure of the wire. This value, then, may be determined from a tensile test taken on the wire at relevant temperatures and strain rates.

11.2. THE TENSILE TEST

11.2.1 Basic description

For the tensile test, a given length of wire, ℓ_0, of cross-sectional area, A_0, is subjected to progressive elongation, usually at a constant rate of elongation. A schematic representation of the starting configuration is shown in Figure 11.1. The data collected are the pulling force, F, and the extent of elongation, $(\ell - \ell_0)$, where ℓ is the length to which the initial section, of length ℓ_0, has been extended at a given stage of the test. Figure 11.1 is schematic, and a formal specimen design, suitable for rod, is presented in Figure 11.2.[64] The reduced section in Figure 11.2 contains the gage length, ℓ_0. In a test to failure, it is hoped that fracture will occur within the gage length, allowing straightforward use of $(\ell - \ell_0)$ throughout the entire test. In the tensile testing of wire, it is usually impractical to use a reduced section, and the gage length is simply placed between the points of gripping. Careful attention must be given to the grip design to minimize the risk of breakage at, or within, the grips. ASTM Standard E8/E8M-08 addresses the tensile testing of wire.[65]

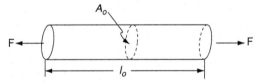

Figure 11.1 Schematic representation of loading in a tensile test.

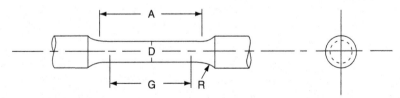

Figure 11.2 Round tensile test design from ASTM Standard E 8/E 8M-08. From *Standard Test Methods for Tension Testing of Metallic Materials*, 2009 *Annual Book of Standards*, Vol. 03.01. ASTM International, Philadelphia, PA, 2009, 64. Copyright held by ASTM International, West Conshohocken, PA, USA.

11.2.2 Equipment

The tensile test may be conducted on an apparatus such as shown in Figure 11.3, with one end of the sample secured to a fixed platen, while the other end is gripped by a moving, screw-driven crosshead. The force is monitored by a load cell attached in series with the specimen, while extension and rate of extension are either projected from the crosshead drive or measured by a device (an "extensometer") attached directly to the specimen gage section. However, wire samples will generally not have a reduced section and may be too thin for the attachment of an extensometer.

11.2.3 Elongation and force

The tensile specimen is generally extended until it breaks, and a generic force-elongation readout is shown in Figure 11.4. The test begins at point A, and between points A and B, the force is very nearly proportional to the elongation. If the test is stopped in this region, the specimen will "spring back" to its original dimensions ($\ell \rightarrow \ell_0$, etc.). The region, A to B, where the elongation is recoverable and proportional to the force, is referred to as the *elastic* region.

At point B, the specimen begins to undergo permanent elongation. If the test is stopped beyond point B, a certain part of the elongation is recovered

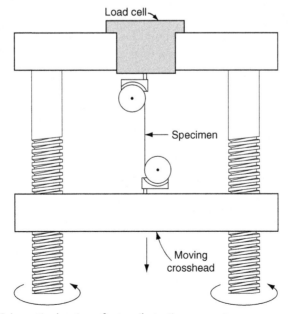

Figure 11.3 Schematic drawing of a tensile testing apparatus.

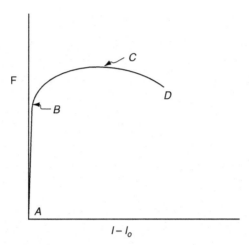

Figure 11.4 Typical pattern of force-elongation data from a tensile test.

with the specimen unloading along a line parallel to the slope of the line from A to B. The recovered elongation is, again, elastic deformation. The permanent or non-recoverable elongation is described as *plastic* deformation.

Beyond point B, the load-bearing capability of the specimen is seen to rise gradually to the point C, at which point a *neck* or local thinning is observed. All further elongation, beyond C, is concentrated within the neck. The force-bearing capability of the specimen steadily diminishes and *fracture* ensues, within the neck, at point D. The neck and the concentration of elongation therein, are illustrated in Figure 11.5.[66]

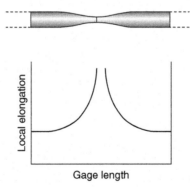

Figure 11.5 Schematic drawing of necked tensile specimen and the variation of elongation along the gage length. From G. E. Dieter, *Mechanical Metallurgy*, Third Edition, McGraw-Hill, Boston, MA, 1986, 294. Copyright held by McGraw-Hill Education, New York, USA.

11.2.4 Stress-strain representation

The data of Figure 11.4 are dependent on specimen size. To generalize the behavior, the size factor is eliminated by converting the force, F, to a stress and the elongation, $(\ell - \ell_0)$, to a strain. As discussed in Sections 4.1.5 and 4.1.6, stress is simply force divided by the cross-sectional area upon which it is acting, whereas strain is the elongation referred to the starting gage length.

There are two common systems used for defining stress and strain in the tensile test: *engineering* stress and strain and *true* stress and strain. It is important to know each system, and to use the system most appropriate to a given technical situation.

Engineering stress, in the tensile test, is defined by the expression

$$\sigma_e = F/A_0, \tag{11.4}$$

where A_0 is the original cross-sectional area of the gage section. Engineering strain, in the tensile test, is given by the expression

$$\varepsilon_e = (\ell - \ell_0)/\ell_0. \tag{11.5}$$

True stress, in the tensile test, is defined by the expression

$$\sigma_t = F/A, \tag{11.6}$$

where A is the instantaneous cross-sectional area of the gage section; that is, as the sample is elongated, it thins (volume can be treated as constant beyond the elastic range) and the true stress reflects the ever-decreasing cross section. True strain, in the tensile test, reflects the continuously changing length of the gage section. By summing up, or integrating, the change in ℓ divided by the instantaneous value of ℓ, or $(d\ell/\ell)$, Equation 11.7 is developed.

$$\varepsilon_t = \ln(\ell/\ell_0) \tag{11.7}$$

Alternatively, this expression can be used,

$$\varepsilon_t = \ln(A_0/A), \tag{11.8}$$

and Equation 11.8 *must* be used within the neck.

The data from Figure 11.4 can then be plotted on an engineering stress-strain basis, as in Figure 11.6a, or on a true stress-strain basis, as in Figure 11.6b. For reference, points A, B, C, and D are indicated again in Figure 11.6a and b. There is little difference in the two stress-strain plots through point B. It is useful to know that point C on the true stress-strain curve, the point of neck formation, is the point where the slope of the curve $(d\sigma/d\varepsilon)$

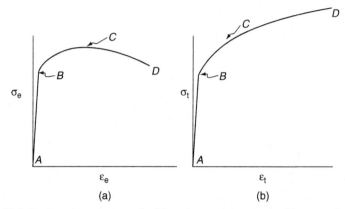

Figure 11.6 Engineering stress-strain (a) and true stress-strain (b) curves based on Figure 11.4.

coincides with the actual value of the true stress. By the time point D is reached, a great deal of change in dimension has occurred, and the values of the two curves are quite different.

11.2.5 Critical points, ranges, and definitions

Stress is proportional to strain in the elastic range from point A to point B, and the proportionality constant is Young's modulus, E; that is, $\sigma = E\varepsilon$ in this range. There are no radial or transverse stresses in the elastic range. There are radial strains, ε_r, however, and $\varepsilon_r = -\nu\varepsilon$, where ν is Poisson's ratio.

The stress at point B in Figure 11.6a and b is generally referred to as the yield strength, and is the initial value of σ_o. It is common practice to use an offset technique to quantify the yield strength, as illustrated in Figure 11.7a. Basically, a line parallel to line AB is drawn from a point on the strain axis, where the strain at that point is called the offset strain. The stress at the point where this line crosses the stress-strain curve is designated as the offset yield strength. Common values of offset strain are 0.2% and 0.02%, although much higher values are sometimes used.

Some wire materials, notably carbon steels, will display complex, "yield point" behavior at yielding, resulting in an upper and lower value of the yield strength, as illustrated in Figure 11.7b. Offset technique is not generally applied to such yield phenomena. A more detailed discussion of yielding behavior in steels is presented in Chapter 14.

The engineering stress value at point C in Figure 11.6 is called the ultimate tensile strength, σ_{eu}. The corresponding true stress value is occasionally used as the ultimate tensile strength, but such use is unconventional and apt to be confusing. The stress values at point D in Figure 11.6 are referred to as the

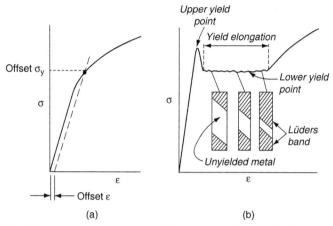

Figure 11.7 Schematic representation of (a) offset yield strength determination and (b) the yield point phenomenon. From G. E. Dieter, *Mechanical Metallurgy*, Third Edition, McGraw-Hill, Boston, MA, USA, 1986, 198. Copyright held by McGraw-Hill Education, New York, USA.

fracture stress. The engineering strain value at point C is called the uniform elongation, ε_{eu}, and the engineering strain value at point D is called the total elongation. The true strain value at point C may be called the uniform true strain, ε_{tu}. True strain values beyond point C must be calculated using Equation 11.8, and the true strain value at D is called the fracture strain. The strain at fracture is often described as the "percent area reduction at fracture," and is defined as

$$\% \text{ area reduction at fracture} = [1 - (A_f/A_0)] \times 100, \qquad (11.9)$$

where A_f is the cross-sectional area at fracture.

As stated in Section 11.1.4, prior to necking, the true stress-true strain curve is literally the effective stress-effective strain curve, or flow curve. And as noted previously, flow curve data are often described by the curve-fitting equation $\sigma = k\varepsilon^N$, where k and N are curve-fitting constants reflecting the material. Since at the point of necking, the slope of the true stress-true strain curve equals the true stress, one can set forth the expression

$$d\sigma/d\varepsilon = N\,k\,\varepsilon^{N-1} = k\varepsilon^N = \sigma \qquad (11.10)$$

and dividing through by $k\varepsilon^{N-1}$, Equation 11.11 is created,

$$N = \varepsilon, \qquad (11.11)$$

and the curve-fitting constant N is often used as a measure of uniform elongation.

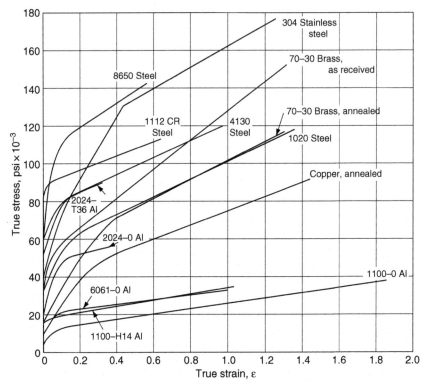

Figure 11.8 Representative flow curves. From *Fundamentals of Deformation Processing*, W. A. Backofen, J. J. Burke, L. F. Coffin, Jr., N. L. Reed and V. Weiss, editors, Syracuse University Press, Syracuse, NY, USA, 1964, 88. Copyright held by Syracuse University Press, Syracuse, NY, USA.

Some representative flow curves are displayed in Figure 11.8.[67] Such data are typically generated at room temperature and at a strain rate in the range of $10^{-3} s^{-1}$. Wire is usually drawn at higher temperatures (due to thermomechanical heating) and much higher strain rates. Figure 11.9 displays flow stress data for aluminum at a range of temperatures and strain rates.[68,69] It is clear that flow stresses for wire and rod drawing modeling must reflect the actual strain rates and temperatures of the drawing process, and tensile testing under such conditions may be impractical, except in an advanced laboratory setting. Fortunately, data are available allowing projection of flow stresses at higher temperatures and stain rates based on room temperature tests run at, for example, a strain rate of $10^{-3} s^{-1}$. Methods and data for such extrapolation for specific metallurgical systems are detailed in Chapters 13, 14, and 15.

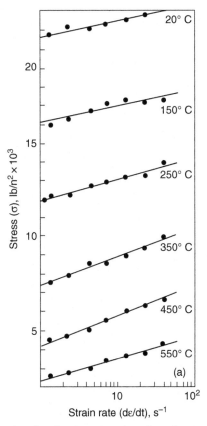

Figure 11.9 Flow stress data for aluminum as a function of temperature and strain rate. From J. F. Alder and V. A. Phillips, *Journal of the Institute of Metals*, 83 (1954–1955) 82. Copyright held by Maney Publishing, London, UK, www.maney.co.uk/.

11.2.6 Approximate determinations of average flow stress

When $\sigma = k\varepsilon^{N}$ flow stress representations are not available, a number of shortcuts to average flow stress representation may be useful. Equation 5.14 presents the average flow stress value in a drawing pass as the average of the flow stresses before and after drawing, or

$$\sigma_a = (\sigma_{00} + \sigma_{01})/2, \qquad (5.14)$$

where the values σ_{00} and σ_{01} would be the yield strengths before and after drawing, at the appropriate temperature and strain rate. If characterization is limited to one tensile test, it is best to approximate σ_a with the ultimate tensile strength of the wire prior to the drawing pass.

11.2.7 Toughness

It is of particular interest that the area under the true stress-true strain curve is the energy per unit volume consumed in deforming the specimen to a given strain. This energy per unit volume is often calculated in projecting process work or power requirements and related values of heating. The total energy per unit volume en route to fracture is called *toughness*. In bulk structural materials, toughness and resistance to fracture are often measured in terms of impact tests or fracture toughness tests. Some bar and rod sizes allow standard impact specimens to be prepared, and research studies have developed fracture toughness specimens appropriate to rather brittle wire materials such as tungsten. However, wire geometry is generally not suitable for such standard testing. Practical aspects of toughness in wire are addressed in Chapter 12.

11.3. THE CRYSTAL PLASTICITY BASIS FOR THE FLOW CURVE

11.3.1 Crystals, polycrystals, and texture

To understand plastic flow it is necessary to consider that the materials drawn are crystalline metals; that is, they are composed of atoms stacked in a repetitive array. For example, the repeating unit, or unit cell, found in copper and aluminum is shown in Figure 11.10. The crystal structure made up of the unit cell "building blocks" shown in Figure 11.10 is called face-centered-cubic (FCC), because a cube is involved with atoms at the centers of each face and at the corners. The geometry of the unit cells may be described with great sophistication: quantitative treatments of planes, directions, spaces among atoms (interstices), and so on. Such crystallographic

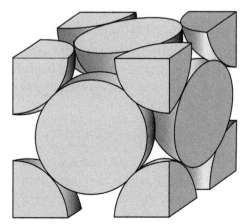

Figure 11.10 Unit cell of the FCC crystal structure.

analysis is beyond the scope of this book. Some physical properties may vary considerably with direction within the unit cell or crystal structure. For example, Young's modulus may be much different in the direction along the cube edges than along the diagonal of the cube. Materials with properties that vary with direction are called *anisotropic*, and materials with properties that do not vary with direction are called *isotropic*.

In some cases, an engineering metal may be composed of just one large crystal, with nearly all of the atoms in positions corresponding to the stacking of unit cells with a specific orientation. More commonly there are many crystals within a piece of metal. Each crystal may have the same structure, but the orientation of the unit cells differs from one crystal to the next. Such metals are called *polycrystals*, and the individual crystals are called *grains*. A cross section of a polycrystalline metal is shown in Figure 11.11.[70] The metal has been sectioned, polished, and chemically etched to reveal the grains and their mutual boundaries. The *grain boundaries* are very thin regions (a few atom diameters thick) where the crystal orientation abruptly changes. It is of interest that fine wires can have very small grain sizes (as small as a few

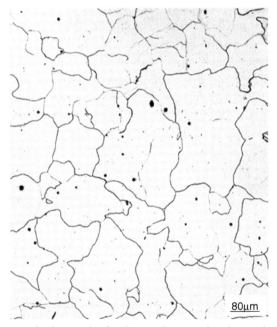

Figure 11.11 An etched section of a low carbon steel polycrystal displaying grain boundaries. From *The Making, Shaping and Treating of Steel, Tenth Edition*, William K. Lankford, Jr., Norman L. Samways, Robert F. Craven, and Harold E. McGannon, Editors, 1985, 1232. Copyright held by Association for Iron & Steel Technology, Warrendale, PA, USA.

micrometers). Grain size can have an important effect on strength, with smaller grain size strengthening metals at lower temperatures and weakening metals at higher temperatures. Grain size also affects the rate of chemical transformation in many metals.

In some cases, the orientations of the many grains in a polycrystal are quite independent from each other. This situation of *random grain orientation* is significant, since the physical properties will be isotropic. In many cases, however, the individual grains are not randomly oriented, and many grains will be oriented close to a *preferred grain orientation*. Such a material is said to have a *texture*. A highly preferred grain orientation, or a "sharp" or "strong" texture, will have properties approaching those of a single crystal, and can have rather anisotropic physical properties.

11.3.2 Slip, dislocations, and plastic flow in crystals

In most cases plastic deformation involves the relative *slip* or *shearing* of certain planes of atoms. The planes may be described rigorously in terms of unit cell geometry, and the deformation is somewhat analogous to the deformation of a deck of cards. The slip occurs when a critical value of stress is reached, and the critical value is relatable to the stress at point B in Figure 11.6a and b.

Slip generally occurs on crystal planes that are the most *closely packed*. In the FCC structure of Figure 11.10, the closest packed planes are planes with normals in the directions of the cube diagonals. Beyond this, slip occurs in close packed directions within the close packed planes. These directions are along lines where the "billiard ball" atoms are in contact with each other. In Figure 11.10, the close packed directions are along the face diagonals. Thus there are only certain types of planes and directions that involve slip. The combination of a *slip plane* and a *slip direction* is called a *slip system*. If the slipping crystal has very smooth, unconstrained surfaces, the slip produces microscopically observable slip bands on the crystal surface. An example of this on an electropolished copper alloy surface is shown in Figure 11.12.[71]

11.3.3 The role of dislocations

Actually, slip cannot occur at the "low" strengths manifested by many metals without the involvement of crystal defects called *dislocations*, which are distortions of crystal structure that extend along a line in the crystal. Figure 11.13 is a schematic illustration of one type of dislocation; namely an edge dislocation, and the manner in which it moves through the crystal to facilitate slip. The slip occurs at a relatively low value of shear stress because

Figure 11.12 Surface of cold worked copper (previously electropolished) showing slip bands. Courtesy of C. Brady, National Bureau of Standards (now the National Institute of Standards and Technology, Gaitherburg, MD, USA).

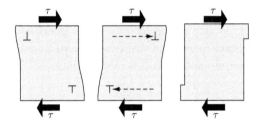

Figure 11.13 Schematic illustration of the motion of an edge dislocation, as it facilitates slip.

the dislocation motion involves the rearrangement of a relatively small number of atoms at a time. If the dislocation was not involved, the entire plane of atoms would have to move at once, and a very large stress would be required. The edge dislocation is just one type of dislocation. Another type is the screw dislocation, and dislocations typically have a mixture of edge and screw character. In any case, slip is the means by which plastic deformation occurs in crystals, and slip occurs through dislocation motion.

11.3.4 Strengthening mechanisms

Anything that interferes with dislocation motion will make slip more difficult and *increase the strength of the crystal*. This is a fundamental principle in the design of many high strength metals, and it provides a basis for understanding work hardening and several other strengthening mechanisms.

Let us consider work hardening. As plastic strain accumulates, the number of dislocations begins to multiply, and the total length of dislocations per unit volume of crystal begins to increase. This increase in dislocation density leads to a certain congestion in dislocation motion; that is, pile-ups and tangles of dislocations occur that increase the stress required for continued dislocation motion (and slip). Thus, the metal becomes stronger as strain accumulates, and the slope of the true stress–strain curve beyond yielding is positive. At lower temperatures (e.g., three-tenths of the melting point), the pile-ups and tangles are not readily dissipated, and the metal can be semi-permanently strengthened by plastic deformation. This strengthening is called cold work.

There are other ways of impeding dislocation motion to achieve increased metal strength. Dissolving one metal in another, or *alloying*, produces what is called solid solution strengthening because the out-of-size and/or out-of-position alloy element atoms interfere with dislocation motion. Sometimes particles or a new or second phase are generated in the alloy for the purpose of immobilizing or impeding dislocations. This approach is called age hardening or precipitation hardening. Additionally, grain boundaries impede dislocation motion and finer grain size (more grain boundary area) usually leads to greater strength, at least at lower temperatures.

11.3.5 Annealing

If a cold worked structure is heated to above six-tenths of the melting point, the dislocation pile-ups and tangles are rapidly eliminated through increased thermal vibrations of the atoms. This occurs through the processes of recovery and recrystallization. Thus, dislocation motion becomes easier, and the strength increase achieved by cold working or work hardening is eliminated. This softening of the worked metal by heating is called *annealing*. The annealed metal can undergo much more strain without fracturing, and has more *ductility* than the cold worked metal. In extended wire drawing operations, periodic annealing may be required to remove cold work and forestall fracture.

While recovery involves the reorganization of dislocations, recrystallization involves the elimination of most of the dislocations, and the establishment of new grain structure. Annealing at rather high temperatures and/or for long times leads to growth of the new grains and a coarse grain structure. Textures or preferred grain orientations developed during cold working often lead to strong, preferred grain orientations in the annealed grain structure. The preferred orientation after annealing may be different than the one developed by cold working, and it is called an *annealing texture*.

11.3.6 Hot work and warm work

If plastic deformation is undertaken above about six-tenths of the melting point (i.e., above the recrystallization temperature range), continuous recovery and/or recrystallization take place with the working, and no appreciable dislocation tangling or work hardening is observed. Working with simultaneous recrystallization is called *hot working*, and can be extensively undertaken without the necessity of annealing. Some metals, like lead, have such low melting points that room temperature is above six-tenths of the melting point, and ordinary handling produces hot working. Some working operations at elevated temperature occur with simultaneous recovery, but not recrystallization. These operations are often referred to as *warm working*.

11.3.7 Other deformation mechanisms

In most cases, plastic deformation involves slip and dislocation motion. However, there are exceptions, and at least three other mechanisms should be mentioned: *twinning, diffusion,* and *grain boundary sliding.* Deformation twinning is a shearing of the crystal to create a "mirror image" across a twinning plane. A schematic illustration is shown in Figure 11.14.[72] The local shear strain is rather large, but the total volume fraction of twinned metal remains small, and only a little overall deformation results. Twinning seems to occur at a critical shear stress that is usually higher than that

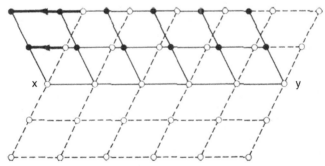

Figure 11.14 Schematic illustration of twinning in a crystal. Dotted lines and open circles represent initial structure, solid lines and solid circles represent twinned structure, and line x-y represents the twin boundary. From D. Hull, *Introduction to Dislocations, 2nd Edition*, Pergamon Press, Oxford, UK, 1975, 27. Copyright held by Elsevier Limited, Oxford, UK.

required to initiate slip. Thus deformation twinning is usually only seen when slip is difficult.

At higher temperatures atoms move among each other by *diffusion*. Such motion can free up blocked dislocations, and the general mass transport by diffusion can produce significant strain.

At lower temperatures, grain boundaries are not regions of weakness, and even increase crystal strength by impeding dislocation motion. However, at temperatures above half the melting point, and with low deformation rates, the grain boundaries may be weaker than the grain interior regions in some metals. Under these conditions, intense shear may occur near the grain boundaries. This phenomenon is often referred to as grain boundary sliding.

11.4. OTHER MECHANICAL TESTS

The tensile test has been discussed previously, and its elements are fundamental to a wide variety of wire tests. Similarly, aspects of hardness testing, compression testing, bend testing, torsion testing, creep testing, and fatigue testing are found in wire mechanical evaluation formats. Basic engineering concepts of these tests are discussed in the next sections. Wire quality control tests and fabrication and service performance simulations are often more complex, even combining these tests modes. A detailed discussion of this is beyond the scope of this chapter, and engineers and technicians are urged to reduce test analysis to the most fundamental format available.

11.5. HARDNESS TESTS

There are several common hardness tests: Rockwell, Brinell, Vickers, Knoop, and variations within these tests. In general, these tests measure the resistance of the metal to an indenter. The indenter may be a ball, a pyramid, or a cone pressed into the metal surface under a given load. A hardness number is calculated from the load and the cross-sectional-area, or the depth, of the indentation. A schematic illustration is presented in Figure 11.15.[73] For example, the Brinell test employs a 10 mm diameter sphere of steel or tungsten carbide, and the Brinell Hardness Number (BHN) is given by the formula:

$$BHN = (2M)/\{(\pi D_i)[D_i-(D_i^2-d_i^2)^{1/2}]\}, \qquad (11.12)$$

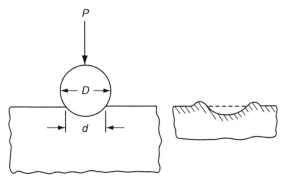

Figure 11.15 Schematic illustration of a Brinell hardness indentation. From S. Kalpakjian, Mechanical Processing of Materials, D. Van Nostrand Company, New York, 167, 36. Copyright held by S. Kalpakjian, Boise, ID, USA.

where M is the mass of the indenter, D_i is the indenter diameter, and d_i is the indentation diameter. The stronger the metal, the smaller d_i and the larger BHN will be. Small loads may be used for soft metals and larger loads for stronger metals. Microscopic and nanoscopic tests are available to resolve phase-to-phase hardness variation, variations in cold work, or near-surface variations. Micro-hardness and nanohardness testing are especially valuable tools for wire evaluation, and may be easily combined with metallographic evaluation.

The primary value of the hardness test lies in the fact that hardness correlates well with strength. Values for BHN can be converted to those of flow stress, at perhaps 5 to 10% strain, based on plasticity analysis. (Equation 11.12 involves mass and not force units, and leads to values such as kg/mm^2. However, $1 \, kg/mm^2$ is equivalent to a stress of 9.81 MPa or 1.42 ksi.) Empirical correlations of hardness with yield strength, and especially tensile strength, are widely available in the literature. Figure 11.16 shows some typical strength correlations.[74]

In carbon steels, hardness correlates very well with ultimate tensile strength. In more rapidly work-hardening metals, such as annealed copper, aluminum, or austenitic stainless steel, the flow stress at 5 to 10% strain is well below the ultimate tensile strength, and carbon steel hardness-tensile strength correlations should not be applied to these other metals as they require their own hardness versus tensile strength correlations.

 11.6. COMPRESSION TESTS

There are contrasts and similarities between tensile and compression tests. In both cases, a gage section is loaded axially, and stress-strain curves can be generated. A typical compression test configuration is shown in

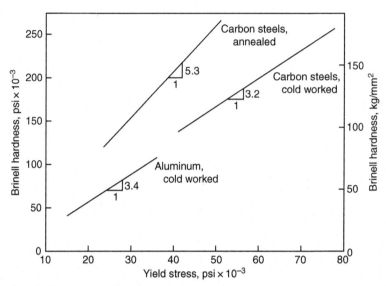

Figure 11.16 Representative strength-hardness correlations. From S. Kalpakjian, *Mechanical Processing of Materials*, D. Van Nostrand Company, New York, 167, 37. Copyright held by S. Kalpakjian, Boise, ID, USA.

Figure 11.17.[75] Since the specimen cross section increases, necking does not occur in compression and the force-elongation curve does not peak. Samples subject to compression may *buckle*; however, and this limits the length-to-diameter ratio of practical specimens to no larger than three (and even less when there is little or no work hardening). Beyond this, the friction between the compression heads and the specimen ends may result in an artificially high load if the length-to-diameter ratio is much less than two. The frictional constraint at the specimen ends produces a barrel-shaped specimen, as illustrated in Figure 11.17, and friction can be reduced with lubricants.

Stresses and strains are calculated in the same manner as in a tensile test. Axial strain will be negative and the compressive stress is mostly considered negative as well. However, many practical compression stress-strain plots use positive values.

11.6.1 Complications of compression testing

Interpretation of compression test results is not straightforward for the following reasons.

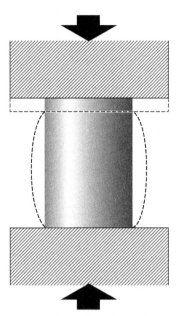

Figure 11.17 Compression test configuration. From H. W. Hayden, W. G. Moffatt and J. Wulff, *The Structure and Properties of Materials, Vol. III, Mechanical Behavior*, John Wiley & Sons, New York, 1965, 10.

The apparent applied stress may reflect friction, as well as flow stress. For example, the axial pressure, P, borne by a cylindrical specimen of axial length h and diameter d is given by

$$P = \sigma_o[1 + (\mu d)/(3h)] = \sigma_o[1 + (\mu)/(3\Delta)], \qquad (11.13)$$

where μ is the coefficient of friction, and Δ is h/d. Thus, instead of P directly measuring the flow stress, σ_o, it overestimates it by the factor $[1 + (\mu d)/(3h)]$ or $[1 + (\mu)/(3\Delta)]$.

The apparent applied stress may reflect the plastic buckling criterion, instead of flow stress. Figure 11.18 is a schematic representation of a compression specimen undergoing plastic buckling. The Euler criterion for such buckling is

$$P = [\pi d/(2h)]^2 (d\sigma/d\varepsilon) = [(\pi/(2\Delta)]^2 (d\sigma/d\varepsilon), \qquad (11.14)$$

where $d\sigma/d\varepsilon$ is the slope of the stress–strain curve. Relative vulnerability to buckling depends heavily on the slope of the stress–strain curve. Work hardening results in a positive slope and buckling resistance. There is no buckling resistance in the absence of work hardening or in the presence of

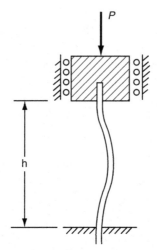

Figure 11.18 Schematic representation of a compression specimen undergoing buckling.

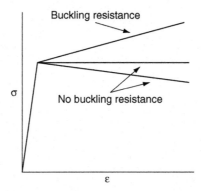

Figure 11.19 Relationships of stress-strain curve slopes to buckling.

work softening (negative work hardening). Figure 11.19 shows a schematic representation of these conditions.

Specimen preparation is somewhat tedious (ends must be parallel).

Test deformation is reversed from that of drawing, and strength may be different from that displayed in drawing. In fact, the reversal of strain will often result in a decrease in flow stress and even work softening.

11.7. BENDING TESTS

Bend testing is widely employed as a simple means of estimating the ductility of the wire surface. Typically, the wire is wrapped around a mandrel of a given radius, D. Sometimes a length of the wire is used as the mandrel. It can be shown that the axial strain, ε, in the bent wire is given by

$$\varepsilon = (2y)/(D + d), \tag{11.15}$$

where y is the distance from the center of the wire (in the plane of bending) and D and d are the mandrel and wire diameters, respectively. The value of y is negative on the concave side of the wire and positive on the convex side. Thus, on the concave side the strain is negative (compressive), and on the convex side the strain is positive (tensile). The maximum absolute value of the strain, ε_{max} , is where y equals d/2, and

$$\varepsilon_{max} = 1/(1 + D/d). \tag{11.16}$$

Axial strain is zero at the wire center, as in Equation 11.15, and the center may be regarded as the "neutral axis."

The bending moment, M_y, associated with yielding and plastic flow of the wire is given by

$$M_y = [(\pi/(32)] \, d^3 \sigma_o. \tag{11.17}$$

Ignoring work hardening, the maximum moment reached during bending, M_L, is $1.7\,M_y$.

Typically, the wire is bent around a mandrel and inspected for surface cracking or gross fracture. The severity of this test can be varied simply by changing the mandrel diameter. In relatively ductile metal, cracking may be somewhat shallow, since the strain decreases proportional to the radial position within the wire. Hence, the degree of cracking defining failure must be carefully established if gross cracking is not expected. For any given "fracture strain," ε_f, a minimum value of (D/d), or of D if the wire diameter is not a variable, exists, and

$$(D/d) = (1/\varepsilon_f) - 1. \tag{11.18}$$

A common bending test is the *wrap test*, as illustrated in Figure 11.20. During this test the wire is simply wrapped around itself, (D/d) is unity, and $\varepsilon_{max} = \frac{1}{2}$ (ignoring torsion). Thus, if a wire or a rod passes a wrap test, then ε_f is greater than one-half.

11.8. TORSION TESTS

Torsion testing involves twisting the wire, either to failure or to some predetermined number of twists, whereupon inspection for surface cracking can be undertaken. The strain of interest is the engineering shear strain, γ,

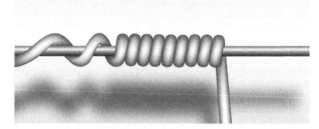

Figure 11.20 Wire configuration in a common bend test called the wrap test. The value of ε_{max} is about one-half.

which involves displacements in the wire cross section plane in the circum-ferential direction. The value of this strain is given by

$$\gamma = 2\pi\, y\, N_t/L, \qquad (11.19)$$

where N_t is the number of full twists, and L is the length of wire twisted. The value of N_t may be usefully expressed in terms of an angle of twist, θ_t, especially if N_t is small, less than one, and so on. In that case,

$$N_t = \theta_t/(2\pi), \quad \text{with } \theta_t \text{ in radians, or} \qquad (11.20)$$

$$N_t = \theta_t/(360), \quad \text{with } \theta_t \text{ in degress.} \qquad (11.21)$$

Then γ is maximum at the wire surface, where y equals d/2, or

$$\gamma_{max} = \pi\, d\, N_t/L. \qquad (11.22)$$

The twisting moment, T_y, associated with yielding and plastic flow of the wire is given by

$$T_y = (0.1134)\, d^3\sigma_o. \qquad (11.23)$$

Ignoring work hardening, the maximum moment reached during torsion, T_L, is $(4/3)\, T_y$.

For any given "fracture strain,", γ_f, a maximum number of twists, N_{tmax}, or a maximum twisting angle exists, and

$$N_{tmax} = L\,\gamma_f/(\pi\, d), \qquad (11.24)$$

where θ_{tmax} can be substituted for N_{tmax}, consistent with Equations 11.20 and 11.21.

A common twist test setup is as illustrated in Figure 11.21.

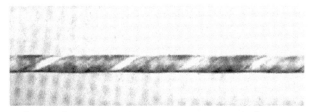

Figure 11.21 Wire configuration in a common twist test. Surface striations indicate that γ_{max} is about two-thirds.

 ## 11.9. CREEP TESTING

In many cases, metals are observed to deform with time, under a constant load or stress, as opposed to simply deforming in conjunction with increased stress. This behavior becomes very important above four-tenths of the melting point, but can be significant over long periods of time at lower temperatures. This time–dependent deformation under constant loading is called *creep*, and tests designed to measure the strain that accumulates with time under constant stress are called creep tests. For example, wire can be loaded in tension and the elongation measured with time. The force may be adjusted, as the cross section changes, to maintain constant stress. The typical accumulation of strain with time is schematically represented in Figure 11.22.[76] After "instantaneous" elastic deformation, and a certain *transient creep* response, an extensive constant strain rate or *steady-state creep* response is seen. Eventually, with internal void formation and necking, a rapid increase in strain is observed and fracture results. The transient range, the steady-state range, and the rapid final range are, respectively, referred to as primary, secondary, and tertiary creep, or as Stage I, Stage II, and Stage III creep. Overall, the creep rate is generally higher with increased stress and increased temperature.

A somewhat simpler test is the *stress rupture* test. In this test, creep behavior is measured by loading a sample at a given temperature, and recording the time to fracture, or the *rupture time*. Typically, a number of these tests are run at varying temperatures and loads, and data are generated as shown in Figure 11.23.[76]

Creep and stress rupture tests thus provide data that indicate the stress and temperature ranges a metal can be exposed to without excessive distortions in a given time at a given stress.

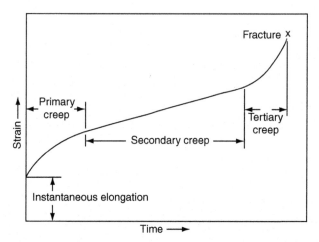

Figure 11.22 Schematic illustration of the accumulation of strain with time in a creep test. From H. W. Hayden, W. G. Moffatt and J. Wulff, *The Structure and Properties of Materials, Vol. III, Mechanical Behavior*, John Wiley & Sons, New York, 1965, 17–18.

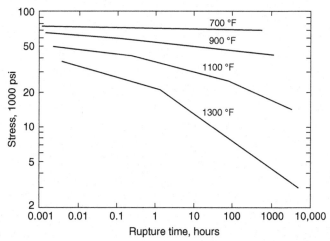

Figure 11.23 Representative stress rupture test data. From H. W. Hayden, W. G. Moffatt and J. Wulff, *The Structure and Properties of Materials, Vol. III, Mechanical Behavior*, John Wiley & Sons, New York, 1965, 17–18.

Metals constrained under a stress established for long-time service performance, as in the cases of a tightened bolt or a crimped wire, will manifest a certain creep rate. The same can be said for metals containing *residual stresses*, established by non-uniform deformation or heating. This creep lowers the established stress by exchanging plastic strain for elastic strain, and the behavior is generally referred to as *stress relaxation*. This stress relaxation can be grossly accelerated at temperatures above four-tenths of the melting point. Thermal processing to achieve this end, and essentially eliminate the stress, is called *stress relief annealing*.

11.10. FATIGUE TESTING

Fatigue is the fracture of a metal by cyclic stressing or straining. Wire fatigue testing can be readily undertaken by reverse bending or torsion cycles. Fatigue testing of wire involving axial loading is limited to the tension range, since buckling would normally occur in compression.

The most efficient method used to determine the high cycle fatigue strength of wire samples is *rotating bending*. With this technique, a bending strain is imposed on a wire sample, and the sample is rotated, *while remaining in the bending plane*. A practical approach involves fixing one end of a length of wire in a freely turning chuck. The length of wire is then bent 180°, with a bending radius of D/2, and the other end is fixed in a driven, rotating chuck. The centerlines of the two chucks and the wire are coplanar, and remain so during testing. As the driven chuck turns, the wire is subjected to cyclic stress and strain from compression to tension, consistent with Equation 11.15. The cycles to fracture are monitored with a revolution counter, or calculated from test time and drive motor speed. A schematic illustration of such a fatigue testing system is shown in Figure 11.24.

11.11. SPRINGBACK TESTING

Elastic springback occurs when stress is released, and it is an important consideration in wire forming and coil winding, as well as in bending and torsion tests. It is important to consider when a wire or rod is bent around a mandrel of a given radius. Upon release of applied stress, the wire or rod will spring back to assume a radius of curvature larger than the mandrel radius. For round wire or rod subject to bending around a mandrel of radius R_0, the springback is described by the formula

$$1/R_1 = 1/R_0 - (3.4)(\sigma_0/E)(1/d), \qquad (11.25)$$

where R_1 is the radius of curvature of the wire after springback.

Lower values of flow stress, σ_o, and higher values of E involve reduced springback. Basic determination of σ_o and E may be complicated by the

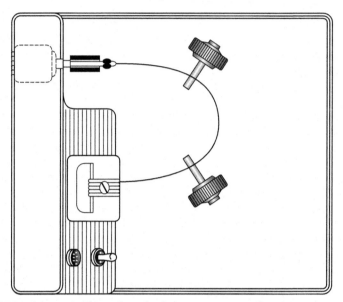

Figure 11.24 Schematic illustration of a fatigue test system for wire.

presence of residual stress. It is often useful to simply measure σ_o/E from the springback test, rather than deal with values of σ_o and E independently.

It is important to realize that the flow stress value involved in Equation 11.25 will be dominated by the flow stress of the near-surface region, since strain is proportional to radial position in the wire, as in Equation 11.15. Therefore springback testing is useful for assessing the mechanical properties of the near-surface region.

If creep occurs during the time that the wire or rod is held on the mandrel at radius of curvature R_0, then springback will be reduced. The *spring elongation test*, widely used to evaluate copper annealing response, involves elements of both elastic springback and time-dependent strain.

▷ 11.12. QUESTIONS AND PROBLEMS

11.12.1 In a simple tensile test, the pulling stress, σ, is the same as the principal stress σ_I. There are no transverse stresses, so σ_{II} and σ_{III} are zero. Use Equation 11.2 to show that yielding occurs when σ equals the effective stress, σ_o.

Answer: Using Equation 11.2, the principal stress values of σ, 0 and 0 can be inserted into the right side of the equation and show that $\sigma = \sigma_o$.

11.12.2 In a simple tensile test, the plastic strain in the pulling direction, ε, is the same as the principal strain ε_1. There are also transverse strains such that ε_{II} and ε_{III} are both equal to $-\varepsilon/2$. Using Equation 11.3, show that ε is equal to the effective strain, ε_o.

Answer: Using Equation 11.3, one can insert the principal strain values of ε, $-\varepsilon/2$, and $-\varepsilon/2$ into the right side of the equation and show that $\varepsilon = \varepsilon_o$.

11.12.3 Equation 11.13 demonstrates that the axial pressure in a compression test is not equal to the flow stress when friction is present. Calculate the P to σ_o ratio for tests with coefficients of friction of 0.05, 0.1, and 0.2 for the cases of $\Delta = 1$ and $\Delta = 3$.

Answer: For the case of $\Delta = 1$, Equation 11.13 predicts P/σ_o ratios of 1.017, 1.033, and 1.067, respectively, for μ values of 0.05, 0.1, and 0.2. For the case of $\Delta = 3$, Equation 11.13 predicts P/σ_o ratios of 1.0056, 1.011, and 1.022, respectively, for μ values of 0.05, 0.1, and 0.2. Notice the contribution of friction to pressure is less for the higher Δ value.

11.12.4 A rather brittle 2 mm diameter wire has a fracture strain of 0.1. What is the smallest mandrel diameter about which this wire can be wrapped without cracking?

Answer: Using Equation 11.18, it can be shown that D/d is nine. Therefore the diameter, D, of the smallest mandrel that can be wrapped without fracture is 18 mm.

11.12.5 A 10 cm length of 2 mm diameter rod must pass a twist test involving 10 turns. The engineering shear strain at fracture is 0.9. Will the rod pass the test?

Answer: Using Equation 11.22, the value of γ_{max} can be calculated as 0.628. Since this is well below 0.9, the rod will pass the test.

11.12.6 A piece of steel rod, 2 mm in diameter, is wrapped about a 5 cm diameter mandrel, and is found to assume a 3 cm radius of curvature upon release. Given that Young's modulus for steel is about 200 GPA, estimate the yield strength of the steel rod.

Answer: Using Equation 11.25, and keeping the radii and diameters straight, one can solve for σ_o as 788 MPa.

CHAPTER TWELVE

Drawability and Breaks

Contents

12.1. PRACTICAL DEFINITIONS

Drawability is the degree to which rod or wire can be reduced in cross section by drawing through successive dies of practical design. This is expressed in apparent true strain, $\varepsilon_t = \ln(A_0/A_1)$. Strictly speaking, such an expression should factor in redundant strain. However, this is rarely done in practice, although practical die designs and pass schedules do involve a moderate level of redundant work. Redundant work will generally reduce the possible drawing reduction, but no adjustment for this is usually made in drawability analysis unless Δ values are very high.

Wire Technology
ISBN 978-0-12-382092-1, DOI: 10.1016/B978-0-12-382092-1.00012-9

When drawing breaks occur at the die exit or at the capstan, the *drawability limit* has been reached. Drawability reflects a given metallurgical condition and *flaw* population. Typical flaws of importance are surface defects (crow's feet, drawn-in "dirt," etc.), and centerline defects (center bursts, porosity from solidification processing, etc.). Drawability may deteriorate with drawing and handling. It may be restored with the removal of cold work by annealing, although it must be understood that annealing will not eliminate flaws.

12.2. MEASURING AND ESTIMATING DRAWABILITY[77]

12.2.1 The role of the tensile test

Drawability may be evaluated with a tensile test, since tensile test fractures develop similar to certain drawing breaks. Figure 12.1 shows a schematic illustration of the manner in which porosity development at flaws or "fracture centers" limits area reduction in a tensile test.[78] With most wire materials, the most significant indicator of drawability from a tensile test is the *area reduction at fracture*. Fracture strain is often expressed as percent area reduction at fracture as shown in Equation 11.9:

$$\% \text{ area reduction at fracture} = [1-(A_f/A_0)] \times 100, \qquad (11.9)$$

where A_f is the cross-sectional area at fracture. However, it is better expressed as a true fracture strain, $\varepsilon_f = \ln(A_0/A_f)$. Very small values of A_f are especially important, and may require scanning electron microscopy (SEM) for meaningful measurement. On the other hand, metals of limited drawability may display quite limited reductions in area.

The most relevant measurements are those undertaken at the temperatures and strain rates of the drawing operation.

12.2.2 The Cockcroft and Latham workability criterion

From a metalworking research perspective, drawability is a special case of the more general concept of *workability*. Workability is the degree to which a material may be plastically deformed prior to fracture. Workability may be quantified by the Cockcroft and Latham fracture criterion.[79] The overall Cockcroft and Latham fracture criterion is given as follows:

$$\int_0^{\varepsilon_f} \sigma_o(\sigma^*/\sigma_o) \, d\varepsilon_o = c, \qquad (12.1)$$

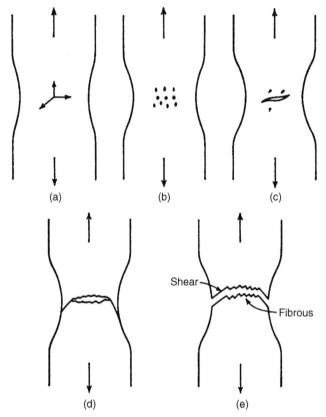

Figure 12.1 Schematic illustration of ductile fracture development in a tensile test. From G. E. Dieter, *Mechanical Metallurgy*, Third Edition, McGraw-Hill, Boston, MA, 1986, 262. Copyright held by McGraw-Hill Education, New York, USA.

where σ_o is effective stress, ε_o is effective strain; (σ^*/σ_o) is a dimensionless stress concentration factor representing the effect of the maximum tensile stress, σ^*; and c is a material constant reflecting workability. It is understood that σ^* can only be positive in the integral, and that for purposes of integration, negative values of σ^* are replaced with a value of zero. Equation 12.1 reduces to

$$\int_0^{\varepsilon_f} \sigma^* \, d\varepsilon_o = c. \tag{12.2}$$

The rationale for this criterion can be explained as follows. First, resistance to fracture is toughness, and toughness can be expressed by the area under the effective stress–effective strain curve, or

$$\text{Toughness} = \int_0^{\varepsilon_f} \sigma_o \, d\varepsilon_o. \qquad (12.3)$$

Second, it is assumed that pore growth, as illustrated in Figure 12.1, will not occur unless tension is present. Thus, "ductile damage" en route to fracture is only caused by the combination of plastic work *and* tension. Thus, Equation 12.3 is modified by the multiplier (σ^*/σ_o), with the further limitation that σ^* can only be positive in the integral, and that for purposes of integration, negative values of σ^* are replaced with a value of zero. Finally, the integration is complete when the fracture strain is reached, and the value of that integral is a measure of workability, equal to c.

Thus, the Cockcroft and Latham material constant, c, is a fundamental index of workability and drawability. It may be evaluated by way of the tensile test, since the true stress and true strain values of the tensile test are related to σ_o and ε_o, or to the effective stress and strain values, as discussed in Section 11.1.4.

Prior to necking, the value of σ_t in a tensile test is in fact σ^*, and the value of ε_t is ε_o. Therefore up to that point the evaluation of the integral in Equation 12.2 is straightforward just like the area under the true stress–true strain curve. Beyond the point of necking, ε_t and ε_o must be evaluated as $\ln(A_0/A_1)$. Moreover, σ^* must reflect the radial tensile stresses in the neck, and must be calculated from a function describing the neck geometry. Figure 12.2 illustrates the distribution of σ^* in the neck cross section for two necking geometries. Therefore, the rigorous evaluation of the integral in Equation 12.2 is rather tedious.

Fortunately, the dominant aspect of the integral in Equation 12.2 is the true strain at fracture, or $\ln(A_0/A_f)$, which can vary widely among materials with diverse workabilities. In contrast, the average value of σ^* can be roughly approximated with the true stress at the point of necking, σ_{tu}, or

$$\sigma^*_{ave} \approx \sigma_{tu} = \sigma_{eu}(1 + \varepsilon_{eu}). \qquad (12.4)$$

Therefore,

$$\int_0^{\varepsilon_f} \sigma^* \, d\varepsilon_o \approx (\sigma^*_{ave}) \ln(A_0/A_f) = \sigma_{eu}(1 + \varepsilon_{eu}) \ln(A_0/A_f) = c. \qquad (12.5)$$

It is important to continue to note the dominant influence of $\ln(A_0/A_f)$ in Equation 12.5. In this context, it may be expedient to use $\ln(A_0/A_f)$ as the relative or comparative measure of workability and drawability.

As a predictor of drawability, $\ln(A_0/A_f)$ is most useful when the condition limiting drawability exists throughout the entire cross section, or when

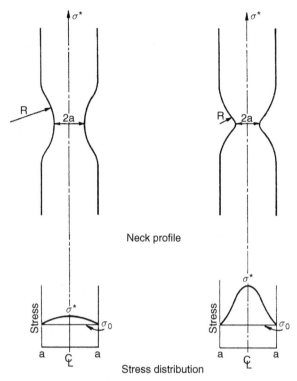

Neck profile

Stress distribution

Figure 12.2 The distribution of the maximum tensile stress, σ^*, in the neck cross section, for two necking geometries. From Thomas A. Kircher, Evaluation and Comparison of Workability for Two Limited-Workability Steels, M.S. Thesis, Rensselaer Polytechnic Institute, 1985.

it is at the centerline. Centerline weakness can result from casting porosity, alloy element concentrations, and newly forming center bursts.

12.2.3 Evaluating workability at the rod surface

When drawability is limited by surface conditions, such as crow's feet, oxide, rolled-in "dirt" and fines, twist tests and bend tests may be better indicators of drawability. This is because twist and bend tests maximize the strain at the rod or wire surface, leading to fracture development and drawability inference at the surface.

12.2.4 Evaluating workability with bending tests

Equation 11.16 leads to the bending fracture strain expression

$$\varepsilon_f \approx 1/[1 + (D/d)_{min}], \tag{12.6}$$

where $(D/d)_{min}$ is the smallest ratio of mandrel diameter to wire diameter that can be used without fracture. Clearly, one can subject wire to bending and wrap tests, infer the fracture strain, and use the fracture strain value as an indicator of drawability.

12.2.5 Evaluating workability with twist tests

Equation 11.22 leads to the twisting fracture strain expression

$$\gamma_f \approx \pi d\, N_{tmax}/L, \qquad (12.7)$$

where N_{tmax} is the maximum number of twists that can be administered to the wire without fracture. Clearly, one can subject wire to twist tests, infer the fracture strain, and use the fracture strain value as an indicator of drawability.

12.2.6 Compression testing and the criterion of Lee and Kuhn[80]

While the problematical aspects of compression testing have been noted in Section 11.6.1, compression tests have been developed for careful analysis of workability at the surface. The pertinent compression test geometry is illustrated in Figure 12.3. The surface strains are described in terms of coordinates in the axial direction, h, and the circumferential direction, s. Thus, upon plastic deformation, the axial compressive strain, ε_z, is $\ln(h/h_0)$

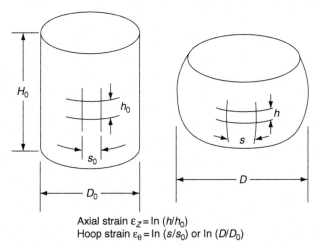

Axial strain $\varepsilon_z = \ln (h/h_0)$
Hoop strain $\varepsilon_\theta = \ln (s/s_0)$ or $\ln (D/D_0)$

Figure 12.3 Geometry for workability testing in compression. From P. W. Lee and H. A. Kuhn, *Workability Testing Techniques*, G. E. Dieter Editor, American Society for Metals, Metals Park, OH, USA, 1984, 49. Copyright held by ASM International, Materials Park, OH.

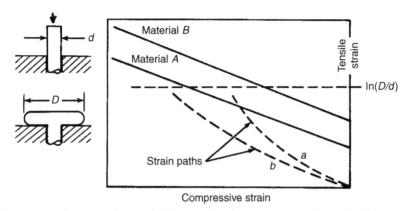

Figure 12.4 Geometry for workability testing in compression. From P. W. Lee and H. A. Kuhn, *Workability Testing Techniques*, G. E. Dieter Editor, American Society for Metals, Metals Park, OH, USA, 1984, 49. Copyright held by ASM International, Materials Park, OH.

and the circumferential tensile strain, ε_θ, is $\ln(s/s_0)$ or $\ln(D/D_0)$, where h_0 and h are starting and as–deformed axial direction dimensions, s_0 and s are starting and as–deformed circumferential direction dimensions, and D_0 and D are starting and as–deformed diameter maximums. Measurement of h and s values may be facilitated with a grid established (such as by etching) on the specimen surface.

Figure 12.4 displays ε_θ versus ε_z relationships for tests a (Material A) and b (Material B), up to a circumferential strain of $\ln(D/d)$. If the surface strains simply reflected uniform deformation, or equal values of circumferential and radial strains, then the strain paths would have followed a straight line relating tensile strain to compressive strain. In general, however, at a certain stage of the compression, the radial strain becomes less than the circumferential strain, and a "flat area" begins to develop on the surface, leading to a condition approaching plane strain with ε_z and ε_θ as the only strains continuing to change. Thus strain paths a and b turn upward away from the straight line. This condition of "plastic instability" leads to fracture once the strain path has departed a critical distance from the straight line or has intersected the limits labeled Material A or Material B.

The fracture conditions just described are defined by the fracture criterion of Lee and Kuhn[80]namely:

$$\varepsilon_\theta + \tfrac{1}{2}\varepsilon_z = q, \tag{12.8}$$

where q is a material property reflecting workability of the surface material. The respective values of q for Material A and Material B are to be found at the intercept of the respective lines with the tensile strain axis.

Various test options exist to establish a database to determine the value q, including that implicit in the diagrams to the left in Figure 12.4.

12.3. CATEGORIZING DRAWING BREAKS

Any time the wire breaks it is a significant event, whether in relation to down time and lost production or to scrap generation and lost product. Moreover, a drawing break signifies that something is wrong with the wire or with the wire processing. In this context, *all drawing breaks should be subjected to real-time scrutiny and saved for analysis.* If nothing else, the operator should register an immediate opinion and related observations, and the wire break ends should be cut off and placed in a labeled envelope. With appropriate microscopic and macroscopic observations, drawing breaks can be categorized as follows.

12.3.1 Category one: Breaks that do not reflect general wire quality, or damage from passage through the die

It is important to single out breaks that have nothing basic to do with the wire quality or with the actual drawing operation. Common examples are

Breaks at obvious cuts or abrasions. A typical example is shown in Figure 12.5.

Breaks at welds. A typical example is shown in Figure 12.6.

Breaks reflecting wire route damage. These may include damage that occurs where a wire crosses over another wire on a capstan, or where abrasion results from contact with worn sheaves and guides.

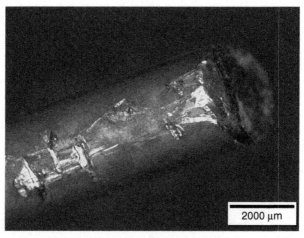

2000 μm

Figure 12.5 Typical example of a wire break at a cut or abrasion.

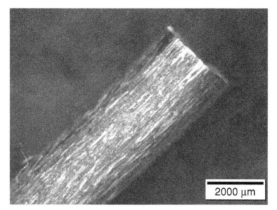

Figure 12.6 Typical example of a wire break at a weld. The longitudinal surface striations were generated in dressing the weld.

12.3.2 Category two: Breaks that primarily reflect mechanical conditions during passage through the die

The breaks in this category cannot be related to wire quality. They reflect solely the plastic flow of the wire in the die and the role of the lubricant. Common examples are

Wire tensile breaks due to a draw stress that exceeds the wire tensile strength. A typical example is shown in Figure 12.7.

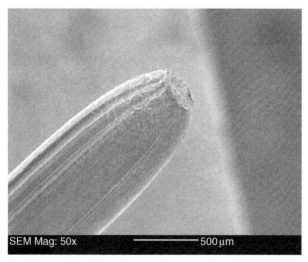

Figure 12.7 Typical example of a wire tensile break. From E. H. Chia and O. J. Tassi, Wire Breaks — Causes and Characteristics, Nonferrous Wire Handbook, Vol. 2, The Wire Association, Inc., Guilford, CT, 1981, 60.

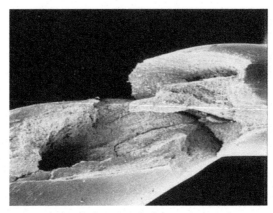

Figure 12.8 Typical example of a wire break due to a center burst. Note "cup" on left, and "core" on right. (Courtesy of Horace Pops)

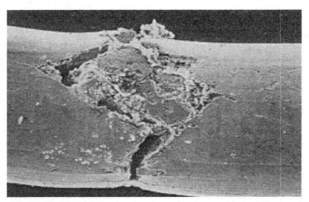

Figure 12.9 Typical example of a wire break due to a surface inclusion. (Courtesy of Horace Pops)

b. Breaks due to center bursts ("cuppy core" breaks). A typical example is shown in Figure 12.8.

c. Breaks due to crow's feet. Such a scenario is shown schematically in Figure 8.16.

12.3.3 Category three: Breaks where metallurgical or microstructural flaws in the wire greatly accelerate development of category two breaks

These breaks reflect the quality of the wire being drawn, and the role of wire flaws is directly evident in the fractography or morphology of the break. Generally such flaws occur at the wire center, where they exacerbate

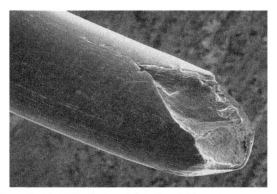

Figure 12.10 Example of an "inclusion absent break" in copper wire, where it is likely that a steel inclusion has fallen out. (Courtesy of Horace Pops)

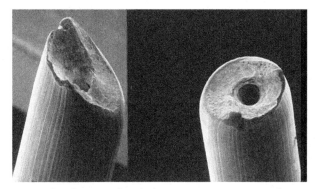

Figure 12.11 Example of an wire break due to macroporosity at the wire centerline. (Courtesy of Horace Pops)

the development of center bursts, or on the wire surface where they complicate friction and surface flow. Common examples are shown in Figures 12.9, 12.10 and 12.11.

 ## 12.4. MECHANICS OF DRAWING BREAKS[81]
12.4.1 The ratio of draw stress to flow stress

In the simplest concept of drawing, a break occurs when the draw stress equals the yield and/or breaking stress of the wire at the die exit. Yielding and breaking are generally associated because of the instability created by the plastic stretching of the wire between dies. Such a break is called a tensile break, since the conditions for failure in a tensile test are largely reproduced.

For practical analyses, we will utilize Equation 5.13:

$$\sigma_d/\sigma_a = \Sigma = [(3.2/\Delta) + 0.9](\alpha + \mu), \qquad (5.13)$$

and take the position that when σ_d/σ_a, or Σ, equals one, the tensile break condition is essentially satisfied. At this point, vibrations, inertial loads, and lubricant fluctuations can be expected to put the draw stress "over the top" as far as yielding and fracture are concerned.

It is important to note that back stress, σ_b, can add significantly to the value of Σ. Equation 5.20 can be modified to display this role:

$$\sigma_d/\sigma_a = \Sigma = [(3.2/\Delta) + 0.9](\alpha + \mu) + \sigma_b[1-(\mu r/\alpha)(1-r)^{-1}]. \quad (12.9)$$

It should also be noted that relatively brittle wires may actually break on contact with the capstan where the tensile bending stress adds to the drawing stress. For a wire of diameter d, and a capstan of diameter D, the bending strain at the wire surface is given by $1/(1 + D/d)$, or about d/D when d is much smaller than D. Such bending tensile strains of 0.1% will add substantially to the local surface stress on the as-drawn wire. Wire crossovers are especially threatening in this regard.

12.4.2 Breaks without obvious flaws

Despite the simplistic criterion of Section 12.4.1, it is observed that breaks become intolerably frequent at values of Σ below one, even in wire without obvious flaws. In laboratory operations, or with certain highly controlled industrial situations, a Σ value as high as 0.9 may be tolerable. It is common in high productivity operations to have a Σ value of 0.7 cited as the practical maximum to avoid widespread flaw growth and frequent breakage.[82]Lower Σ maxima will be required in the face of a discernable flaw population.

Equation 5.13 can be used to project maximum reductions consistent with a Σ value of 0.7 for given values of die semi-angle and coefficient of friction. Such reductions are listed in Table 12.1. It is apparent in Table 12.1 that there is only a small effect of die angle on maximum drawing reduction. This observation must be tempered, however, with the understanding that α and μ are not necessarily independent. Die angle reductions can improve lubrication under thick film conditions, and can increase friction when lubrication is marginal.

Friction plays a major role in determining the maximum reduction, however. Table 12.1 indicates that thick film lubrication conditions (μ in the range of 0.03) may permit drawing reductions as high as 40% without frequent breaks. However, bright drawing conditions (μ in the

Table 12.1 Maximum reductions, as a function of die semi-angle and friction coefficient, for a draw-stress-to-average-flow-stress ratio of 0.7.

Die Semi-Angle (°)	Friction Coefficient	Maximum Reduction (%)
4	0.03	41
4	0.10	25
4	0.15	18
6	0.03	43
6	0.10	28
6	0.15	22
8	0.03	43%
8	0.10	30
8	0.15	23
10	0.03	42
10	0.10	30
10	0.15	24

range of 0.10) and marginal lubrication (μ in the range of 0.15) restrict drawing reductions to the 25–30% and 18–24% ranges, respectively.

In this context, a sudden increase in drawing break frequency may reflect increased friction, especially if no flaw population is apparent. As discussed in Chapter 8, such a deterioration in lubrication should be confirmable by microscopy.

12.4.3 Breaks in the presence of obvious flaws

The Σ limit of 0.7, in the absence of obvious flaws, reflects the fact that ductile fracture mechanics models predict the growth of even very small flaws at Σ values above this level. Very small inclusions are largely unavoidable, and even natural, such as copper oxides in ETP copper (see Chapter 13). Moreover, the initiation of center bursts and crow's feet leads directly to a developable flaw. Poor quality wire and poor drawing practice can present flaw sizes that grow at Σ levels well below 0.7. Such flaws include cuts, abrasions, and weld deterioration.

When flaw cross-sectional area, A_{fl}, becomes, for example, within an order of magnitude of the wire cross section, A, it is useful to examine the *net section stress*, σ_{ns}. The net section stress is the stress value obtained by dividing the force, F, by the cross-sectional area $(A - A_{fl})$ that remains when the flaw area is subtracted from the nominal wire cross section $(\pi d^2/4)$. That is,

$$\sigma_{ns} = F/(A - A_{fl}) = (F/A)[A/(A - A_{fl})] = \sigma_d/(1 - A_{fl}/A). \quad (12.10)$$

As an example, if a flaw were 20% of the cross-sectional area of the wire, then $\sigma_{ns} = \sigma_d/(1 - 0.2) = 1.25 \ \sigma_d$.

Since it has been hypothesized that any flaw will grow when Σ or (σ_d/σ_a) exceeds 0.7, we can take the position that any flaw will grow when σ_{ns}/σ_a exceeds 0.7. If we replace σ_a with (σ_d/Σ), flaw growth is predicted when $\Sigma(\sigma_{ns}/\sigma_d)$ is at least 0.7. Replacing the stresses with force and area values, one obtains

$$(A_{fl}/A) = 1-\Sigma/(0.7), \tag{12.11}$$

where (A_{fl}/A) is the relative flaw size that is predicted to grow rapidly at a given Σ or (σ_d/σ_a) level. As a check for consistency, we can note that if the flaw is infinitesimal, and (A_{fl}/A) is zero, then the given value of Σ for flaw growth is 0.7, which is the (σ_d/σ_a) ratio at which we have said that even infinitesimal flaws will grow.

12.4.4 The case of ultra-fine wire drawing[83]

The production of ultra-fine wire (diameters the order of 0.02 mm) is grossly inhibited by the threat of breaks, and the associated loss of very-high-value-added wire. A number of measures are introduced to reduce breaks including screening of redraw rod; decrease in per pass reduction; reduction in drawing speed; use of carefully matched, low angle dies; rigorous lubricant maintenance and control; and intensive drawing machine oversight. Metzler has summarized these measures.[84]

In this context, the break frequency can be been modeled as:

$$B/L \approx A_{fl} \ N_{fl} \ J \ \Sigma, \tag{12.12}$$

where B/L is the number of breaks per unit length of wire, N_{fl} is the number of flaws per unit volume, and J is a fracture index inversely related to wire toughness. Common wire drawing experience suggests that J has a value in the range of four for metal of high toughness, and a value in the range of eight for average toughness. Much higher J values represent brittle wire. Ultra-fine copper wire drawing practice expectations indicate, by way of Equation 12.12, that a cubic meter of very high drawability copper rod should contain and/or develop no more than the order of fifty drawing-break-related flaws.

It is worth noting that Equation 12.12 seems to imply that B/L is independent of wire diameter. This would seem unreasonable given the great increase in break frequency often observed with continued drawing to finer sizes. This increased B/L is accounted for by increases in the values of J, N_{fl}, and A_{fl} with progressive drawing reduction.

Although Equation 12.12 is presented for application to the case of ultra-fine-wire drawing, it can be usefully applied to heavier gage drawing practice. The roles of Σ and A_{fl} remain pertinent, even if N_{fl} requirements are less strenuous.

12.5. THE GENERATION OF "FINES"

During most drawing operations, the surface of the wire undergoes "wear" (wire wear as opposed to die wear), and small pieces of the wire (called "fines") flake off. Such behavior is included in this chapter, since it represents a local form of wire fracture. Figure 12.12 shows an example of an incipient fine in the act of emerging from a copper wire surface.[85]

The fines are involved with surface quality, since their formation leaves rough areas, and since the fines may be pressed into the drawn wire surface. Fines compromise lubrication and metal flow by clogging dies and by chemically reacting with the lubricant.

Gross fine development may reflect poor rod surface quality. Fine development can also reflect factors such as die angle and die alignment. Figure 12.13 shows a relationship of copper fine development to die angle, indicating that an intermediate, 16° included die angle minimizes fine development.[85] Such intermediate die angles are the norm in copper

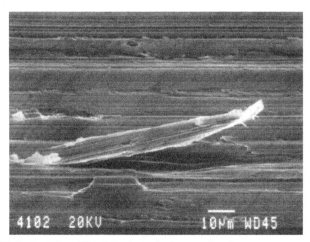

Figure 12.12 An incipient "fine," in the act of emerging from a copper wire surface. From G. J. Baker, Workpiece Wear Mechanisms in the Drawing of Copper Wire, Ph.D. Thesis, Rensselaer Polytechnic Institute, 1994.

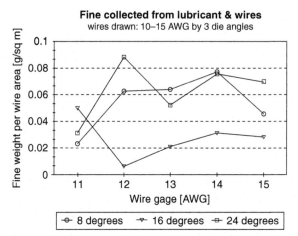

Figure 12.13 Relationship of copper "fine" development to drawing die angle. From G. J. Baker, Workpiece Wear Mechanisms in the Drawing of Copper Wire, Ph.D. Thesis, Rensselaer Polytechnic Institute, 1994.

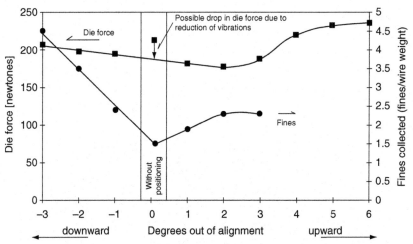

Figure 12.14 Data showing the minimization of copper "fine" development with good die alignment. From G. Baker and H. Pops, *Some New Concepts in Drawing Analysis of Copper Wire, Metallurgy, Processing and Applications of Metal Wires*, H. G. Paris and D. K. Kim (Eds.), The Minerals, Metals and Materials Society, Warrendale, PA, 1996, 29. Copyright held by The Minerals, Metals and Materials Society, Warrendale, PA, USA.

drawing, despite the general advantages often cited for low angle, low Δ drawing passes. Figure 12.14 shows a minimization of fine development with good die alignment.[86]

12.6. QUESTIONS AND PROBLEMS

12.6.1 A rod displays an ultimate tensile strength of 200 MPa, a uniform elongation of 17%, and an area reduction at fracture of 45%. Evaluate the workability constant, c, for the Cockcroft and Latham criterion.

Answer: Equation 12.5 can be used. For a 45% reduction, $\ln(A_0/A_f) = 0.60$, therefore the value of c is 140 MPa.

12.6.2 A low-flaw-content wire is drawn with a die with a 12° included angle. It is found that the maximum practical reduction, without excessive breaks, is 30%. Estimate the coefficient of friction.

Answer: Equation 5.13 can be used with the assumption that σ_d/σ_a can be 0.7. Putting in the values (do not forget the semi-angle) leads to a coefficient of friction of 0.089.

12.6.3 A 2 mm diameter wire has consistent flaws the size of 0.5 mm. What is the largest ratio of drawing stress to average flow stress that can be taken without probable breakage?

Answer: Equation 12.11 can be used, and the ratio (A_{fl}/A) can be approximated by the square of the diameter ratio. On this basis, σ_d/σ_a is estimated at 0.66. This is a high value and is probably only reasonable for a smooth-surfaced pore. Lower figures would probably result with solid, rough-surfaced inclusions.

12.6.4 Consider that the wire in Problem 12.6.3 is quite tough, but that the number of flaws considered to be present is 1000 per m³. How many breaks per unit length can be expected with respective drawing-stress-to-average-flow-stress ratios of 0.5, 0.4, and 0.3?

Answer: Using Equation 12.12, estimating A_{fl} at $(0.5\,\text{mm})^2$ and inserting values, the relation $B/L = 10^{-3} (\sigma_d/\sigma_a)$ is calculated. The values of B/L are 0.0005, 0.0004, and 0.0003 for the respective (σ_d/σ_a) values of 0.5, 0.4, and 0.3. These values correspond to 2000, 2500, and 3000 meters per break.

CHAPTER THIRTEEN

Relevant Aspects of Copper and Copper Alloy Metallurgy

Contents

Wire Technology
ISBN 978-0-12-382092-1, DOI: 10.1016/B978-0-12-382092-1.00013-0

13.1. IMPORTANT PROPERTIES OF COPPER

13.1.1 Conductivity

Copper has outstanding electrical and thermal conductivity, brought about in both cases by electron transfer. In its pure form, copper has a room temperature electrical conductivity of, or somewhat above, 0.5800×10^5 $(\Omega \text{ cm})^{-1}$. Only silver exceeds this value (by about 5%). A conductivity of $0.5800 \times 10^5 (\Omega \text{ cm})^{-1}$ corresponds to that of the International Annealed Copper Standard (IACS). The IACS is defined as a resistivity of $1.7241 \, \mu\Omega$ cm. Resistivity is the inverse of conductivity, and resistivity measurements are often used in copper electrical specifications.

As noted in Table 6.1, the thermal conductivity of copper is $3.85 \, \text{J}/(\text{cm s} \, °\text{C})$, and silver's thermal conductivity is about 8.7% higher.

Copper conductivity is reduced with increased temperature with cold work and alloy content, as discussed in Section 13.6.

13.1.2 Corrosion resistance

Copper and many of its alloys have a very useful level of corrosion resistance. In the galvanic series, their corrosion resistance is below that of noble metals, graphite, titanium, silver, passive stainless steel, and certain nickel alloys, but above that of active nickel, stainless steels, tin, lead, cast irons, carbon steels, and aluminum, zinc, and magnesium systems. The galvanic series represents relative activities in sea water.

13.1.3 Mechanical properties

Copper and most of its alloys have a face-centered cubic (FCC) crystal structure and correspondingly high levels of ductility, toughness, and form-ability. Strength may vary from soft to moderately strong with significant work-hardening response. In this context copper displays outstanding drawability.

13.2. PRIMARY PROCESSING

While the primary processing of copper from its ores is an immense subject complicated by environmental considerations and the availability of scrap, a brief review of the basics may be highly useful for the readers of this book. The basics of two major methods of primary copper production will be discussed, providing a transition into the finish processing of the higher purity, conductor grades. For a comprehensive treatment of this subject, the reader is referred to the work of Biswas and Davenport.[87]

13.2.1 Pyrometallurgy

Roughly 90% of world production of copper comes about by this technology. An ore source of least 1% copper content must be available, with copper mostly in a form equivalent to Cu_2S, Cu_2O, and CuO. The initial processing may involve *calcining* (partial roasting) with reactions such as:

$$2ZnS + 3O_2 \rightarrow 2ZnO + 2SO_2. \tag{13.1}$$

The first major step is *smelting*, or melting with a separation of the charge with silica (sand?). This results in two components, *matte* and *slag*. The matte contains copper and is composed of Cu_2S, FeS, Au, and Ag, among other items of interest. The slag involves silica reactions such as:

$$FeO + SiO_2 \rightarrow FeSiO_3. \tag{13.2}$$

The next operation is called *converting*, where the matte (molten) is blown with O_2 to preferentially oxidize impurity sulfides. Representative chemical reactions are

$$2FeS + 3O_2 \rightarrow 2FeO + 2SO_2 + \underline{Heat} \tag{13.3}$$

$$Cu_2O + FeS \rightarrow Cu_2S + FeO \tag{13.4}$$

$$Cu_2S + O_2 \rightarrow 2Cu + SO_2 + \underline{Heat.} \tag{13.5}$$

The product of Equation 13.5 is the next stage of refinement called blister copper, and it generally contains precious metals as well as copper.

Blister copper is subjected to *fire refining*. This involves blowing the copper with air to oxidize reactive metal impurities to slag. Next, the remaining Cu_2O is reduced by "poling," where wood may be added as a source of CO and H_2, such that:

$$Cu_2O + CO \rightarrow 2Cu + CO_2. \tag{13.6}$$

$$Cu_2O + H_2 \rightarrow 2Cu + H_2O. \tag{13.7}$$

Enough Cu_2O generally remains to require a "deoxidation practice." One approach is to add phosphorous (in the form of Cu-15%P) to remove oxygen as P_2O_5. The resultant product of 99.5% copper is called phosphorous deoxidized copper. Electrical conductivity is lowered, however, by the deoxidant (see Table 13.3).

Alternatively, the fire-refined product may be electrolytically refined. In this electrically driven process, the fire-refined copper is the anode, and highly refined copper is plated out at the cathode. One product of this process is electrolytic tough-pitch (ETP) copper. This grade retains an oxygen content of approximately 300–400 ppm. In this way, electrical conductivity is optimized by tying up solute as oxide (Fe becomes FeO, etc.).

It is a major issue that ETP copper is vulnerable to "blistering" (not to be confused with blister copper cited previously) in the presence of H_2, such as in welding or certain heat-treating atmospheres. The basic chemical reaction is

$$Cu_2O + H_2 \rightarrow 2Cu + H_2O(steam). \qquad (13.8)$$

This problem may be avoided with the use of "oxygen-free" copper grades, with conductivity optimized by achieving 99.9+% purity copper.

The metallurgical product forms that emerge from pyrometallurgical copper processing may be melted for processing into bar, rod, plate, strip, and so on.

13.2.2 Hydrometallurgy and solvent extraction

While this technology produces only about 10% of the world copper production, it is capable of producing very high purity copper, and it is thought to be more environmentally friendly than the pyrometallurgical approach.

Solvent extraction (SX) works most efficiently with copper oxide bearing ores (although sulfide procedures exist). The oxide ore is leached with dilute sulfuric acid, as per

$$CuO + H_2SO_4 \rightarrow Cu^{++} + SO_4^{--} + H_2O. \qquad (13.9)$$

The product, on the right-hand side of Equation 13.9, is called *pregnant leach solution* (PLS).

For concentration and purification purposes, PLS is vigorously mixed with an ion-exchange "organic," RH, as represented below:

$$2RH + Cu^{++} + SO_4^{--} + \text{impurities} \rightarrow R_2Cu + 2H^+ + SO_4^{--} + \text{impurities}.$$
$$(13.10)$$

Hydrogen in the ion-exchange organic is replaced with Cu, with the H appearing as H^+ in the product. Thus, the right-hand-side product contains "loaded organic," R_2Cu, and *dilute* sulfuric acid called raffinate. Now the loaded organic and raffinate on the right-hand side of Equation 13.10 are immiscible and have different specific gravities. Hence, gravity separates the organic (top) from the raffinate (bottom), allowing the organic to be readily drawn off.

Next, the loaded organic is vigorously mixed with *concentrated* sulfuric acid and *stripped* of its copper using Equation 13.11:

$$2H^+ + SO_4^{--} + R_2Cu \rightarrow 2RH + Cu^{++} + SO_4^{--}. \tag{13.11}$$

Note that Equation 13.11 is simply the reverse of Equation 13.10, where the reversal is caused because Equation 13.11 involves concentrated sulfuric acid as opposed to the dilute sulfuric acid in Equation 13.10. Moreover, the concentrated copper sulfate on the right-hand side of Equation 13.11 is, as in Equation 13.10, easily separated from the organic.

Finally, the enriched copper sulfate from the right-hand side of Equation 13.11 is subjected to an *electrowinning* process. In this electrochemical process, the anode is an inert metal (lead, etc.) and the cathode can be copper or is often stainless steel. The electrolyte is the enriched copper sulfate, and copper from the electrolyte is caused to plate out on the cathode. The anode reaction produces oxygen.

The overall hydrometallurgical process just described is often called the solvent extraction-electrowinning process (SX-EW). The product cathode copper may be directly used as material for continuous casting or for other downstream copper processes and applications.

13.2.3 Casting of semi-finished product

As previously noted, the variously refined coppers resulting from the pyro- and hydrometallurgical processes are cast into forms suitable for the manufacture of rod and other forms. Recycled scrap may be cast this way as well. The rod can then be drawn into wire. Historically, copper was cast into discrete ingots and worked down to rod form. In modern times, the vast majority of copper is continuously cast into rod form. A schematic illustration of one such system for copper rod casting is displayed in Figure 13.1.[88] Molten copper is fed into a moving mold involving parallel steel belts held in tension, as the top and bottom of the mold. The mold sides are formed by chains of rectangular blocks. The copper solidifies into a continuous "bar". The bar is then shaped, rolled, pickled and coiled, as discussed in Chapter 17.

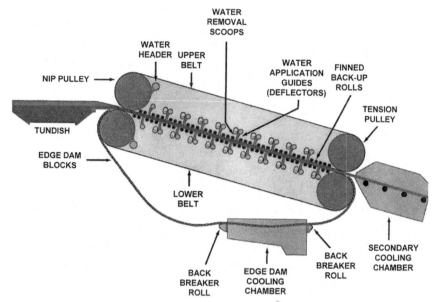

Figure 13.1 Schematic representation of Hazelett® Twin-Belt Casting Machine (Courtesy of Hazelett Strip Casting Corporation).

 13.3. CRYSTAL STRUCTURE, GRAINS, TEXTURE, ANISOTROPY, AND SPRINGBACK

13.3.1 Polycrystalline copper

The basic crystal structure of copper is FCC, as illustrated in Figure 11.10. Four important directions in the FCC unit cell are indicated in Figure 13.2: along the cube edge <100>, along the face diagonal <110>, along the cube diagonal <111>, and from the corner to the face center <112>. The atomic stacking density is different in each of these directions and many physical properties vary according to these (and other) directions. Such property variation is called *anisotropy*.

Most metal workpieces are made of many crystals, and if these crystals are randomly oriented, physical properties may not vary with direction in the workpiece. Such property uniformity is called *isotropy*.

Wire drawing deformation, however, creates *preferred grain orientations*, or *texture*. In this context, wire properties vary with direction, and, as the texture varies in orientation or in non-randomness, the properties in the axial direction (along the wire length) vary. Strong drawing textures may be passed on through the anneal process, although the actual preferred orientation

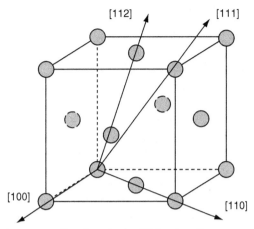

Figure 13.2 Four important directions in the FCC unit cell.

Table 13.1 Effect of copper wire process history on the preferred grain orientation or texture along the wire axis

Process History	Dominant Texture
Cold drawn beyond a cumulative reduction of 99%	<111>
Cold drawn beyond a cumulative reduction of 99% and annealed	<100>
Cold drawn to a cumulative reduction of 90%	<111>, <100>
Cold drawn to a cumulative reduction of 90% and annealed	<111>, <112>, <100>

may change. Table 13.1 summarizes some preferred orientations for copper wire, as a function of process history.

Young's modulus, E, is one of the properties affected by crystal orientation, and hence by texture, as implied by Table 13.2.

This variation in Young's modulus, in turn, affects springback. Springback response was described with Equation 11.25

$$1/R_1 = 1/R_0 - (3.4)(\sigma_0/E)(1/d). \qquad (11.25)$$

Table 13.2 Effect of grain orientation on Young's modulus for copper

Texture	Young's Modulus (GPa)
Random	115
<111>	190
<100>	67

Inspection of Equation 11.25 reveals that higher values of E reduce springback. The very high cold drawing reductions (>99%) leading to a <100> annealing texture result in increased as-annealed springback. Lower springback is generally desired for coil winding applications, and, where low springback is required, such high reductions are avoided by annealing at intermediate stages of cold drawing. A thorough final anneal will promote low springback by reducing the flow stress, σ_o.

Processing en route to low springback response is confounded somewhat by the fact that the high speed drawing of copper can involve substantial heating and recrystallization during the "cold" drawing process.

13.4. FLOW STRESS, COLD WORKING, AND ANNEALING

13.4.1 The flow stress of copper

The flow stress of copper is very sensitive to impurity levels, cold work, and grain size. Ultra-high purity copper may have a yield strength in the range of 0.35 MPa. On the other hand, cold worked electrical conductor copper commonly displays tensile strengths over 200 MPa. Handbooks list strength values, and strength values are specified for various copper product forms; however, a baseline strength or flow stress characterization for copper requires laboratory testing.

Flow stress, as a function of effective strain, can be fitted to and projected from the equation $\sigma_o = k\varepsilon_o^N$. For example, the annealed copper data from Figure 11.8 can be reasonably described by Equation 13.12, with σ_o in MPa:

$$\sigma_o = (525)\varepsilon_o^{0.47}. \tag{13.12}$$

Figure 13.3 displays data for cold worked copper[19] that corresponds to Equation 13.13, again with σ_o in MPa:

$$\sigma_o = (420)\varepsilon_o^{0.12}. \tag{13.13}$$

In addition to the obvious dependence on strain, copper flow stress is also dependent on temperature and strain rate. The effect of strain rate, $d\varepsilon_o/dt$, based on various data in the literature, is reasonably expressed by Equation 13.14:

$$\sigma_o = G(d\varepsilon_o/dt)^{0.012}, \tag{13.14}$$

where G is the strain rate strength coefficient. The effect of temperature is reasonably expressed by Equation 13.15:

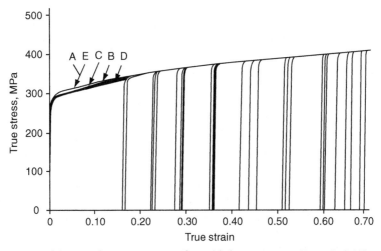

Figure 13.3 Multi-pass flow stress curve for cold drawn copper. From J. G. Wistreich, *Proceedings of the Institution of Mechanical Engineers*, 169 (1955) 657. Copyright held by Professional Engineering Publishing, London, UK.

$$\sigma_o = \sigma_{To} + (-0.45)(T-T_o), \qquad (13.15)$$

where σ_{To} is the flow stress in MPa at the reference temperature T_o, in °C.

Cold working of copper generally increases yield strength, tensile strength, and flow stress. It also decreases ductility and increases electrical resistance. These effects can be largely removed by *annealing*, or heating the copper to a given temperature for a given time. Figure 13.4 displays some typical data, revealing the effect of annealing time and temperature on the tensile strength of ETP copper wire.[89]

It is useful to combine the effects of time and temperature in data plots such as Figure 13.4, since increased time has the same qualitative effect as increased temperature. The combined effects of times and temperatures for simple recrystallization anneals can be expressed by an *annealing index*, I, where

$$I = Log_{10}(t) - Q/[(2.303)RT], \qquad (13.16)$$

where t is time (seconds), T is °K, R is the gas constant or 8.31 J/(mole ° K), and Q is an activation energy. A useful value of Q in copper data analysis is 120.5 kJ/mole. Figures 13.5 and 13.6 show copper annealing responses as a function of I.[89]

Plots such as those of Figures 13.5 and 13.6 allow direct comparison of the effects of longer time anneals (with "box" or "bell" furnaces) to those of

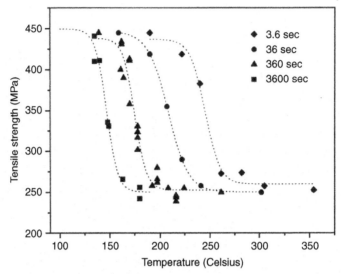

Figure 13.4 Strength versus temperature annealing curves for 12 AWG ETP copper with a cold worked area reduction of 93%. From F. F. KRAFT, Analysis of In-Line Systems for Rapid Annealing, Ph.D. Thesis, Rensselaer Polytechnic Institute, 1994.

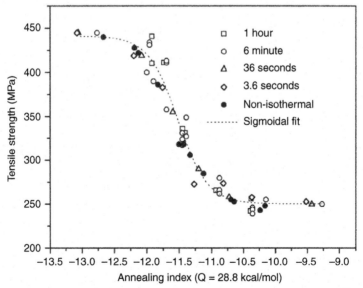

Figure 13.5 Master strength versus annealing index curve for 12 AWG ETP copper with a cold worked area reduction of 93%. From F. F. KRAFT, Analysis of In-Line Systems for Rapid Annealing, Ph.D. Thesis, Rensselaer Polytechnic Institute, 1994.

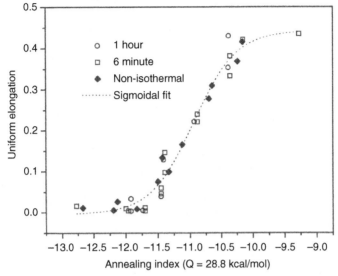

Figure 13.6 Master uniform elongation versus annealing index curve for 12 AWG ETP copper with a cold worked area reduction of 93%. From F. F. KRAFT, Analysis of In-Line Systems for Rapid Annealing, Ph.D. Thesis, Rensselaer Polytechnic Institute, 1994.

much faster, higher temperature "in-line" annealing systems. Figure 13.7 shows a schematic diagram of an in-line electrical resistance annealer, suitable for rapid copper annealing.[90]

As-annealed copper has a characteristic grain size, and as-annealed copper flow stress is a function of that grain size. Flow stress dependence on grain size is commonly expressed by the Hall-Petch relation, or

$$\sigma_o = \sigma_g + k_g d_g^{-1/2}, \qquad (13.17)$$

where σ_g is the flow stress for a single crystal, k_g is a grain size strength coefficient, and d_g is the grain diameter. The value of k_g for copper and simple copper alloys may be in the range of $12\,MPa\,mm^{1/2}$.

13.5. SOLID SOLUTIONS AND PHASES

Alloying elements, processing additions (such as for deoxidation), and "foreign" elements in general may occupy a position in the copper crystal structure. Such elements may replace a copper atom, in which case they may be called *substitutional*, or they may occupy normally open spaces among copper atoms, in which case they are called *interstitial*. These elements are

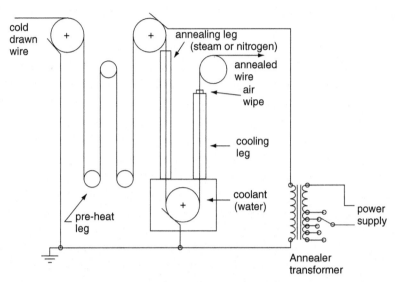

Figure 13.7 Schematic diagram of an in-line electrical resistance annealer. From F. F. KRAFT, Analysis of In-Line Systems for Rapid Annealing, Ph.D. Thesis, Rensselaer Polytechnic Institute, 1994.

said to be soluble in copper, and copper containing this type of solute is called a *solid solution.*

These solid solutions have compositions called *solubility limits,* beyond which a given solute atom will no longer dissolve, but will participate in the formation of a *new phase.* The *stable* relations among composition, temperature, and phases may be presented in a *phase diagram.* Figure 13.8 presents a phase diagram pertinent to the addition or presence of silver in copper.[91] The right-hand portion of the diagram contains a region where a copper solid solution, labeled (β), is present. This phase exists all the way to the melting range, but has hardly any width below about 400 °C. The line labeled "Solubility limit of Ag in Cu" represents the compositions beyond which increasing silver content will not dissolve in the copper, but will form a second phase with a very high silver content. At temperatures well below 400 °C it may be said, from a practical point of view, that silver is insoluble in copper. In contrast, at approximately 779 °C eight weight percent silver can be present in a solid solution of copper and silver. As this solution is slowly cooled, a silver-rich phase will *precipitate,* creating a two-phase, silver-rich structure distributed in the otherwise copper-rich *matrix phase.* If the solution is rapidly cooled, there may not be adequate time for formation of a stable amount of the precipitate phase. In such cases, the matrix phase, or

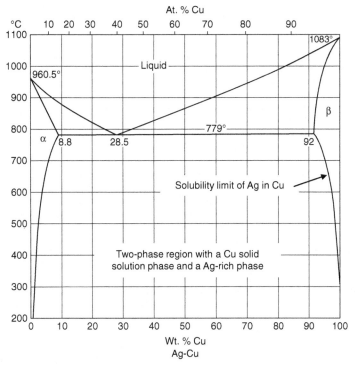

Figure 13.8 Silver-copper phase diagram. From *Metals Reference Book*, C. J. Smithells, editor, Butterworths, London, UK, 1976, 377. Copyright held by Elsevier Limited, Oxford, UK.

solvent, contains solute beyond the stable solubility limit, and is called a *supersaturated solid solution*.

Solid solutions are generally stronger than the base-metal in pure form. Figure 13.9 displays the effect of nickel, as a solute element, on the strength of copper.[92] Copper and nickel can be fully dissolved in one another for any composition, and no second phase will be present below the melting range.

13.6. FACTORS AFFECTING CONDUCTIVITY/RESISTIVITY

The electrical resistivity, ρ_e, of copper and metals in general has three sources that can vary: those associated with dissolved impurity elements, with thermal vibrations, and with crystal defects (dislocations, vacancies, etc.). It is common to express these contributions using Equation 13.18.

$$\rho_e = \rho_{eo} + \rho_T + \rho_i + \rho_d, \qquad (13.18)$$

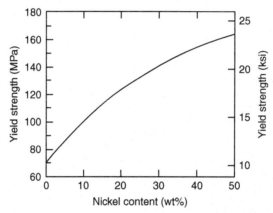

Figure 13.9 The effect of nickel, as an alloying element, on the strength of copper. From W. D. Callister, Jr. and D. G. Rethwisch, *Fundamentals of Materials Science and Engineering*, John Wiley & Sons, New York, 2008, 259. Reprinted with permission of John Wiley & Sons, Inc. Copyright held by John Wiley & Sons, Inc.

where ρ_{eo} is the intrinsic resistivity of the crystal, ρ_T is the contribution of temperature, ρ_i is the contribution of impurities, and ρ_d is the contribution of crystal defects. Figure 13.10 shows data for copper and copper-nickel alloys that illustrate these contributions.[93] The contribution of defects, by way of plastic deformation, is shown in the middle of the figure where the effect of deformation on the resistivity of Cu-1.2 at% Ni is indicated by an offset labeled ρ_d. Otherwise, the differences in the resistivity of the alloys at absolute zero reflect the "impurity" effect of the nickel additions, and the slopes of the five lines reflect the effect of temperature. It should be noted that copper is *not* a superconductor, and it still exhibits some resistivity at absolute zero.

Table 13.3 lists the effects of 0.2 wt% additions of several elements on copper otherwise rated at 100% IACS.

 13.7. DILUTE COPPER ALLOYS

This section addresses major copper alloys with low alloy content, hence the term "dilute alloy."

13.7.1 Oxygen-free copper

The concept of these forms of copper is that high conductivity is achieved by limiting essentially all solute material. Conductivity requirements are generally above 101% IACS. The basic "alloy" is called "oxygen-free electronic copper" and is designated C10100 in the UNS system. It is

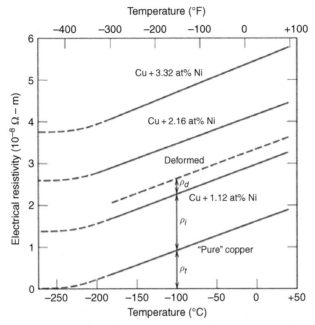

Figure 13.10 Data showing the effect of temperature, alloy content, and deformation on the resistivity of copper. From W. D. Callister, Jr. and D. G. Rethwisch, *Fundamentals of Materials Science and Engineering*, John Wiley & Sons, New York, 2008, 470. Reprinted with permission of John Wiley & Sons, Inc. Copyright held by John Wiley & Sons, Inc.

Table 13.3 Conductivity levels in % IACS for copper alloyed with 0.2 wt% additions of several elements

Added Element at 0.2 %	Conductivity, %IACS
Silver	99
Cadmium	98.6
Zinc	98
Tin	90
Nickel	87
Aluminum	77
Beryllium	64
Arsenic	61
Iron	52
Silicon	47.5
Phosphorous	34

electrolytically refined and must contain at least 99.99% copper. A closely related electrolytically refined alloy is "oxygen-free copper" designated C10200, which may contain a small silver addition to increase the recrystallization temperature and refine grain size. Finally, "oxygen-free, extra-low phosphorous copper," designated C10300, may contain some silver and includes a small amount of phosphorous for deoxidation purposes. Alloy C10300 has an electrical conductivity of 99% IACS.

13.7.2 ETP copper

The concept of this alloy was discussed briefly in Section 13.2.1. The UNS designation is C11000. Although it is electrolytically refined, this grade retains an oxygen content of 300–400 ppm. In this way, electrical conductivity is optimized by the tying up of solute as oxide (Fe becomes FeO, etc.).

The majority of the oxygen in ETP copper forms copper oxide; however, and a brief discussion of the role of oxygen in copper is warranted. The solubility of oxygen in solid copper is extremely limited at most temperatures, leading to the formation of Cu_2O. It does increase near the melting range, reaching a value as high as 0.0075 %, as shown in the small portion of the copper-oxygen phase diagram shown in Figure 13.11.[94] Rapid cooling from this range can lead to supersaturation, with an unstable amount of oxygen in solution, and this phenomenon may affect certain states of copper production. However, oxygen can be regarded as an alloying element in ETP copper used to remove solute elements that would increase resistivity if left in solid solution. The presence of the oxides does not materially affect conductivity.

As noted in Section 13.2.1, ETP copper is vulnerable to "blistering" in the presence of H_2, such as in welding or certain heat-treating atmospheres, and the basic chemical reaction is

$$Cu_2O + H_2 \rightarrow 2Cu + H_2O(steam). \qquad (13.8)$$

In addition to causing vulnerability to blistering, the copper oxides are sites for ductile fracture, and the drawability of ETP copper, although high, is less than that of oxygen-free copper.

The composition limit for ETP copper is 99.90% copper, and the IACS conductivity is at least 100%.

13.7.3 Phosphorous-deoxidized copper

As noted in Section 13.2.1, phosphorous is highly useful as a deoxidizing agent in pyrometallurgically processed copper. However, not all of the phosphorous forms oxide. As much as 1.7% phosphorous can be

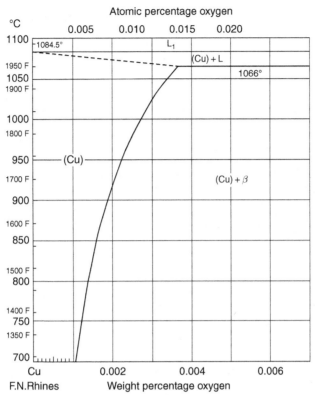

Figure 13.11 A portion of the copper-oxygen phase diagram. From D. T. Hawkins and R. Hultgren, *Phase Diagrams of Binary Alloy Systems, Metals Handbook*, Vol. 8, 8th Edition, American Society for Metals, Metals park, OH, 1973, 295. Copyright held by ASM International, Materials Park, OH.

dissolved in copper near 700°C, although the solubility drops steadily with decreasing temperature. Even so, phosphorous-deoxidized copper can contain as much as 0.04% phosphorous at ambient temperature, thus reducing the conductivity as much as 25%. Thus, phosphorous-deoxidized copper is not generally used for electrical conductor applications.

Phosphorous-deoxidized copper is widely used in hardware applications. It does not present the risk of blistering found with ETP copper, and is suitable for a wide range of joining procedures. Its composition is 99.90% copper. The basic UNS designations are C12000 and C12200, with respective phosphorous levels of 0.008 and 0.02%. Designation 12100 contains a small silver addition.

13.7.4 General dilute alloy and impurity concepts

As indicated above, silver in amounts up to 0.1% may be added to raise recrystallization temperature, refine grain size, and increase creep strength with relatively little decrease in conductivity. Cadmium, in amounts up to 1.0%, may be added to copper to increase strength with relatively little decrease in conductivity. Arsenic, in amounts up to 0.5%, may be added to copper to increase recrystallization temperature and increase high temperature strength.

Lead, in amounts up to approximately 2.0%, may be added to copper to improve machinability. Lead is insoluble in copper and lead "globules" in the microstructure reduce ductile fracture energy, greatly assisting a wide variety of machining and cutting operations. In a similar manner bismuth improves copper machinability. Tellurium additions are reported to improve machinability.

The reduction in ductility presented by alloying with lead to improve machinability may be a major nuisance in some applications. Antimony may reduce ductility in copper, and selenium and tellurium are detrimental to weldability.

13.8. HIGH ALLOY SYSTEMS

13.8.1 Brasses (Cu-Zn alloys)

Figure 13.12 displays the copper–zinc diagram.[95] The so-called *alpha brass* alloys involve the large copper solid solution phase field, labeled α, that runs from pure copper up to over 30% zinc. So-called *beta brass* alloys involve the region around 50% zinc, labeled β. Many beta brass alloys involve mixed (α + β) phase fields.

Zinc is an extremely valuable alloying addition to copper. It increases both strength and ductility (a rare combination). Moreover, zinc additions to copper lower material cost, because zinc is less expensive than copper. On the other hand, zinc does lower corrosion resistance somewhat, with brasses just below copper on the galvanic series.

13.8.2 Alpha brasses

The addition of zinc to copper in the alpha phase field provides solution strengthening. It also increases work hardening, thus inhibiting necking and leading to increased tensile elongation, and its workability is very good.

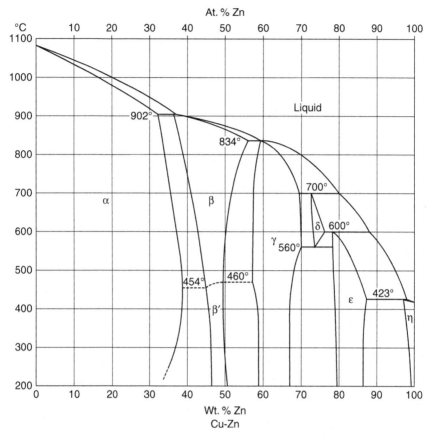

Figure 13.12 Copper-zinc phase diagram. From *Metals Reference Book*, C. J. Smithells, editor, Butterworths, London, UK, 1976, 597. Copyright held by Elsevier Limited, Oxford, UK.

The principal corrosion phenomena are those of stress corrosion cracking, especially in the cold worked state, and "de-zincification." Stress corrosion cracking involves spontaneous cracking in the presence of a stress, such as a residual stress after cold working, or an applied stress, such as in a fastener. The cracking occurs in a critical environment, and ammonia is a common contributor to brass stress corrosion cracking. De-zincification involves the selective leaching of zinc from the brass alloy, leaving a weak, porous structure.

The classic alpha brass alloy is UNS designation C26000, containing about 30% zinc. This alloy is often referred to as "cartridge brass."

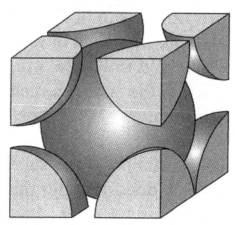

Figure 13.13 BCC unit cell.

13.8.3 Beta brasses

Beta brasses involve the β phase field in Figure 13.12. While the alpha phase has an FCC structure like pure copper, beta brass can have two different crystal structures. At higher temperatures, and upon rapid cooling, the copper and zinc atoms are randomly mixed in a body-centered–cubic (BCC) array, such as shown in Figure 13.13. This condition of random positioning of copper and zinc is called *disordered*. Below about 460°C, the stable crystal structure involves one atom type at the unit cell center and the other type at the cell corners. (This involves a total of one atom each of copper and zinc, consistent with 50 at% of each element.) This structure is called *ordered*. The ordered structure is relatively strong, but with limited ductility. On the other hand, the disordered, higher temperature structure has excellent hot workability.

Many beta brass products involve combinations of alpha grains and beta grains. It is often important to avoid a continuous beta phase, since this phase has less corrosion resistance and can provide a path for corrosive attack.

The most common beta brass alloys are the "free machining" type, which have lead added. The classic beta brass alloy is UNS designation C36000, containing about 35.5% zinc and 3% lead. This alloy is often referred to as "free machining brass."

13.8.4 Tin bronzes

Tin bronzes are alloys of copper and tin, with the FCC structure similar to pure copper. In that sense they are analogous to alpha brass, only with tin as the alloying element instead of zinc. One difference, however, is that tin is relatively expensive. The tin additions greatly increase strength, and yet the alloys remain workable.

Truly historic, "bronze age" alloys contained about 10% tin. Coinage alloys, often called "copper," involve about 5% tin (or combinations of tin and zinc). A basic alloy of technological importance is UNS designation C51000, containing 5% tin, with a small amount of phosphorous. This alloy is commonly called "phosphor bronze" and has many important electrical connector applications.

13.8.5 Aluminum bronzes

Aluminum bronzes are alloys of copper and aluminum, again with the FCC structure. Aluminum additions greatly increase strength, especially high temperature strength, as well as wear resistance. Moreover, aluminum additions provide oxidation resistance and color aesthetics. Classic alloys are UNS designation C60800 with 5 % aluminum that is suitable for cold forming applications, and UNS designation C61000 with 8 % aluminum that requires hot forming.

13.8.6 Copper-nickel alloys

As noted previously copper and nickel are mutually soluble at all compositions. Thus, there is one continuous FCC phase field. Common classical alloys are various coinage metals, "monel," and "nickel silver".

The common coinage alloy, generally called "nickel," is UNS designation C71300 and contains 25% nickel. Monel corresponds to UNS designation C71500 with 30% nickel. Nickel silver, another misnomer, corresponds to UNS designation C77000 with 18% nickel together with a 27% zinc addition.

13.9. BERYLLIUM COPPER, A PRECIPITATION STRENGTHENING (HARDENING) ALLOY

The concept of precipitation was introduced in Section 13.5. In early stages of precipitation, the precipitate phase can exert considerable strain on the surrounding matrix. Dislocations interact with the strain, dislocation motion and slip are made more difficult, and the alloy is said to be precipitation strengthened.

The general procedure for precipitation strengthening involves heating the alloy to a temperature where only one phase exists, and where the alloying element(s) is dissolved in that phase. In the simplest case, the alloy can be rapidly cooled, leading to supersaturation of the alloying element(s). Upon heating to an intermediate temperature, the precipitate emerges, and at a critical degree of precipitation substantial strengthening may be achieved. This subsequent heating process is often called "aging."

In any event, "beryllium copper" is an important alloy involved in precipitation strengthening, often in conjunction with cold work. Typical beryllium content is 2%. An instructive part of the copper–beryllium phase diagram is shown in Figure 13.14.[96] The copper–2% beryllium alloy may be heated to a temperature above 800°C in the FCC alpha phase field, with all

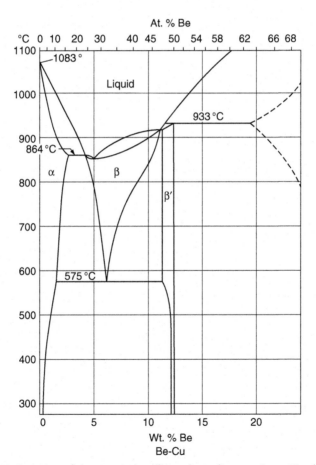

Figure 13.14 A portion of the copper-beryllium phase diagram. From *Metals Reference Book*, C. J. Smithells, editor, Butterworths, London, UK, 1976, 489. Copyright held by Elsevier Limited, Oxford, UK.

of the beryllium dissolved in the copper. This alloy can be rapidly cooled and aged below 300°C. Industrial processing of high strength beryllium copper alloys involves proprietary practices that include combinations of cold work and aging.

A classic beryllium copper alloy is UNS designation C17200, containing 1.9% beryllium with 0.2% cobalt.

13.10. QUESTIONS AND PROBLEMS

13.10.1 Why is the oxygen content (a few hundred ppm) so important for ETP copper?

Answer: Some of the oxygen reacts with solute elements to form oxides. The oxides have little effect on resistivity, whereas the solute increases resistivity. Therefore, the presence of oxygen in the "alloy" keeps the resistivity low and the conductivity high.

13.10.2 Why do some copper grades contain phosphorous? Why are such grades unsuitable for electrical conductor applications?

Answer: The phosphorous is used to deoxidize the copper during final stages of primary processing. Unfortunately, phosphorous decreases electrical conductivity (see Table 13.3).

13.10.3 A welded copper joint is discovered to be full of small holes. Provide an explanation based on information given in this chapter.

Answer: The copper may be ETP copper. Hydrogen may have been present and reacted with the copper oxides to form water in the form of steam (see Equation 13.8). The steam blows open the sites of the copper oxides, leading to blisters. To avoid this, a phosphorous-deoxidized grade or an oxygen-free grade can be used.

13.10.4 A copper annealing cycle runs for an hour at 175°C. What is the value of the annealing index, I, for this cycle? Suppose the annealing temperature was raised to 200°C. How much shorter would the cycle be?

Answer: Using Equation 13.16 and converting to seconds (3600) and °K (448), I can be calculated to be about -10.5. Using this value along with 473 °K, one can solve for time, which is about 660 seconds, or 11 minutes.

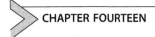
Relevant Aspects of Carbon and Low Alloy Steel Metallurgy

Contents

Wire Technology
ISBN 978-0-12-382092-1, DOI: 10.1016/B978-0-12-382092-1.00014-2

14.1. IMPORTANT PROPERTIES OF STEEL

Steel is the world's most used metal. Its role in engineering and technology is based on two major factors, cost and range of strength. Steel is relatively inexpensive, except for higher alloyed forms. Moreover, steel may be prepared and processed with strength ranging over two orders of magnitude, for example, from 30 MPa to over 3000 MPa. Steel is also available in ranges of toughness adequate for major structural applications.

These advantages are particularly pertinent for the case of *carbon steels*, which are simple alloys of iron and carbon, usually with small manganese additions. Processing flexibility can be increased with small additions of chromium, molybdenum, vanadium, and so forth. These alloys are called *low alloy steels*. This chapter will generally address carbon and low alloy steels.

A major disadvantage of steel is its vulnerability to corrosion ("rusting"). It lies below copper and copper alloys in the galvanic series. This vulnerability to corrosion is substantially mitigated with large chromium additions, and these alloys are called *stainless steels*. Stainless steels are discussed in Chapter 15 of this book. Of course, steel corrosion protection is provided by various coatings, and that subject is addressed in Chapter 16.

Performance in cutting and machine tool applications is greatly enhanced with large alloying additions of a number of elements, forming the basis for alloys called *tool steels*. Tool steels will be discussed in Chapter 15.

 14.2. PRIMARY PROCESSING

The reservations expressed in Section 13.1 for copper processing also apply to steel. The processing of iron from its ores, and the making of steel are immense subjects, complicated by environmental considerations and the availability of scrap. For a comprehensive review of steel basics, the reader is referred to the *Making, Shaping and Treating of Steel*.[97]

14.2.1 Reduction of iron ore

Iron ores are basically iron oxides, and the first step in the overall production process for steel occurs in a *blast furnace*. The charge to the blast furnace consists of iron ore, coke (carbonaceous material), and limestone (mainly calcium carbonate). Preheated air is blown through the charge. Under steady-state operations, the following reactions occur. Near the air blast source, at the base of the furnace, coke burns completely, as per

$$C + O_2 \rightarrow CO_2 + heat. \qquad (14.1)$$

Higher up in the furnace, carbon dioxide is reduced by hot coke, as per

$$CO_2 + C \rightarrow 2CO - heat. \qquad (14.2)$$

Despite the loss of heat in Equation 14.2, a high temperature is retained and carbon monoxide reduces the iron ore. A representative equation is

$$Fe_2O_3 + 3CO \rightarrow 2Fe + 3CO_2. \qquad (14.3)$$

As all of this is going on, waste material from the ore (silica, among many other substances) is converted into *slag* (silicates, etc.) using the following equations:

$$CaCO_3 \rightarrow CaO + CO_2; \qquad (14.4)$$

$$2CaO + SiO_2 \rightarrow Ca_2SiO_4. \qquad (14.5)$$

On a semi-continuous basis, iron ("pig iron") and slag are drawn off, and charge is added. The blast furnace may run continuously for years.

14.2.2 Steelmaking processes

The pig iron contains about 3 to 4.5 % carbon. Although these carbon levels are excellent precursors for *cast iron*, they are much too high for steel. Hence, steel must be processed for carbon reduction in "steelmaking" operations. Such operations have undergone many improvements over the last century and a half. Contemporary technologies are dominated by the *basic oxygen process* and by *electric arc processes*.

14.2.3 The basic oxygen process

This process involves the charging of molten pig iron and steel scrap into a vessel or "converter." A "lance" is inserted into the top of the converter, and oxygen is blown into the liquid. The oxygen reacts with the iron to form an iron oxide, which reacts with carbon, as per

$$FeO + C \rightarrow Fe + CO. \qquad (14.6)$$

Lime, or other "slag-forming fluxes," are added. The overall reactions are self-sustaining, and carbon, manganese, silicon, sulfur, and phosphorous levels are rather quickly reduced, as shown in Figure 14.1.[98]

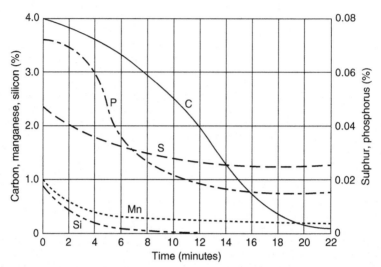

Figure 14.1 Reduction of carbon, phosphorous, sulfur, manganese, and silicon levels during basic oxygen steel making. From W. F. *Smith, Structure and Properties of Engineering Alloys*, Second Edition, McGraw-Hill, Inc., New York, 1993, 86. Copyright held by McGraw-Hill Education, New York, USA.

Variations on the basic oxygen process exist, including the blowing of oxygen from the bottom of the converter, and blowing a mixture of argon and nitrogen from the bottom, to provide stirring of the charge.

14.2.4 Electric-arc processing

In this technology, a cold charge of steel scrap is placed in a vessel, and an electric arc is struck between graphite electrodes and the scrap. The scrap is then melted. This process is especially useful where supplies of scrap are readily available. It also involves comparatively low capital cost, and provides good temperature control.

Electric arc processing may be necessary where easily oxidized elements such as chromium, tungsten, and molybdenum are to be added to alloy steels. Such alloy additions may be excessively oxidized in the oxygen-blowing practices of the basic oxygen process, resulting in costly losses of these additions. In the electric arc process, special slag covers protect against oxidation of alloying elements, as well as lowering phosphorous and sulfur levels.

14.2.5 Ladle metallurgy

Superior and more efficient refining can be achieved by transferring steel from basic-oxygen and electric-arc furnaces to a *ladle*. Ladle processing provides superior temperature and compositional homogenization in the molten steel. Precise deoxidation practices can be achieved with aluminum additions, and alloying additions can be made with increased precision. Cover slags can be employed to reduce sulfur and inclusions can also be floated into the slag and removed. The morphology of residual sulfide and oxide inclusions can be managed by additions of calcium and rare earths.

14.2.6 Vacuum degassing

Additional refinements can be achieved with ladle practice by way of *vacuum degassing*. In the most widely used process, a vacuum degassing system is placed over the ladle with two "legs" or "snorkels" extending into the molten steel. One snorkel introduces argon gas into the steel to force the metal into the degassing system through the other snorkel. Oxygen is introduced in the degassing system to reduce carbon in the steel with carbon monoxide formation. The vacuum degassed steel is returned to the ladle and a new cycle of degassing is initiated. Vacuum degassing practice can achieve very low carbon and hydrogen contents.

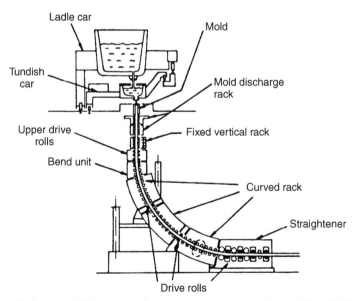

Figure 14.2 Schematic illustration of the continuous casting of steel. From W. F. *Smith, Structure and Properties of Engineering Alloys*, Second Edition, McGraw-Hill, Inc., New York, 1993, 89. Copyright held by McGraw-Hill Education, New York, USA.

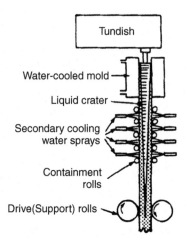

Figure 14.3 Schematic illustration of the mold region in Figure 14.2. From W. F. *Smith, Structure and Properties of Engineering Alloys,* Second Edition, McGraw-Hill, Inc., New York, 1993, 90. Copyright held by McGraw-Hill Education, New York, USA.

14.2.7 Casting practices

While individual ingots may be cast for certain steels, the majority of steel production is continuously cast. A detailed treatment of the continuous casting of steel is outside the scope of this book. However, Figures 14.2 and 14.3 provide illustrative details.[99,100] Rolling of the continuous cast product is discussed in Chapter 17.

 14.3. THE IRON-IRON CARBIDE PHASE DIAGRAM

14.3.1 The phases of iron

The iron-rich end of the iron-iron carbide phase diagram is displayed in Figure 14.4.[101] Despite the relative complexity of this diagram, it is a good and necessary place to start the discussion of carbon steel. The left-hand side of the diagram represents *pure iron*. The left-hand side of the diagram, illustrates that up to 912°C iron exists as a phase called *ferrite*, or α iron. From 910 to 1400°C, iron exists as a phase called *austenite*, or γ iron. Finally, between 1400°C and the melting point at 1535°C, iron exists as a phase called *delta ferrite*, or δ iron.

The right-hand side of the iron-iron carbide phase diagram involves the phase iron carbide, or Fe_3C, often called *cementite*. Cementite is a hard, brittle material with a complex crystal structure. The carbon in Fe_3C is not

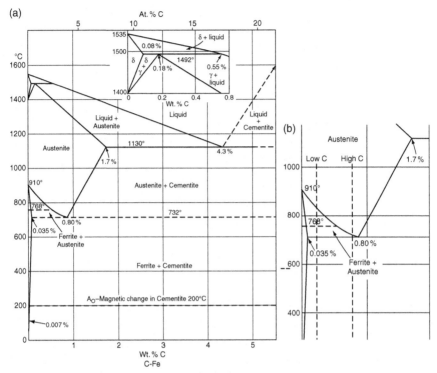

Figure 14.4 (a) Iron-rich end of the iron-iron carbide phase diagram, (b) enlarged portion of (a). From *Metals Reference Book*, C. J. Smithells, editor, Butterworths, London, UK, 1976, 510. Copyright held by Elsevier Limited, Oxford, UK.

as stable as pure carbon or graphite in the iron–carbon system. Thus, strictly speaking Figure 14.4 is not a stable phase diagram. However, for practical purposes it will be considered stable for this discussion. It should be noted that the portion of the phase diagram between ferrite and iron carbide, and below 732°C, consists of a two-phase combination of ferrite and cementite.

14.3.2 Crystal structure and the solubility of carbon

Now ferrite and delta ferrite have a body-centered-cubic (BCC) structure, as shown in Figure 13.13, and shown again for convenience in Figure 14.5. These two forms of iron can largely be regarded as the same phase, albeit separated by a temperature gap. The austenitic phase has a face-centered-cubic (FCC) structure, as shown in Figure 11.10, and again for convenience in Figure 14.6. A major difference between these two crystalline forms of

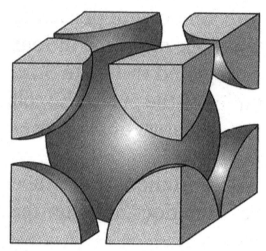

Figure 14.5 The BCC structure, as per ferrite and delta ferrite.

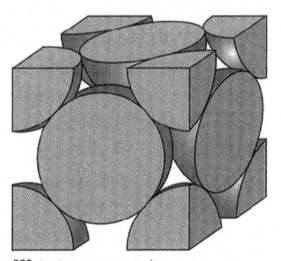

Figure 14.6 The FCC structure, as per austenite.

iron is their ability to accommodate carbon as a solute. Austenite has spacious locations for carbon atoms at the cell center, and at the centers of the cell edges, as illustrated in Figure 14.7. On the phase diagram it is clear that at 1130°C austenite can contain 1.7% carbon as solute, well above the carbon levels of steels. Ferrite, while less dense than austenite, has no such locations for carbon solute, and the maximum solubility of carbon in α ferrite is only 0.035% at 732°C. At carbon levels beyond this small solubility limit, the stable phases are ferrite (nearly pure iron) and cementite (or austenite).

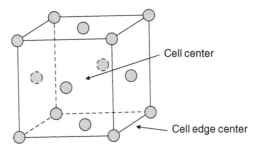

Figure 14.7 Spacious locations for carbon atoms in the FCC structure.

 14.4. AUSTENITE DECOMPOSITION

14.4.1 Transformation of austenite upon cooling

The significance of the points discussed in Section 14.3.2 is that the majority of carbon steel processing scenarios involve, at one or more stages, heating the steel to the austenite range and cooling the steel below 732°C, thus precipitating some combination of ferrite and iron carbide. This 732°C transformation temperature is called the *eutectoid* temperature. A range of possibilities for these *austenite decomposition products* is shown in Figure 14.8.

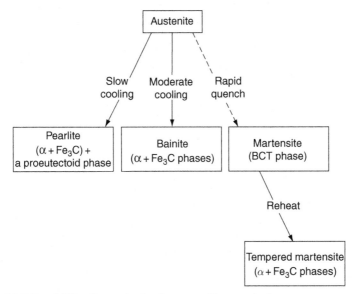

Figure 14.8 Possibilities for austenite decomposition products.

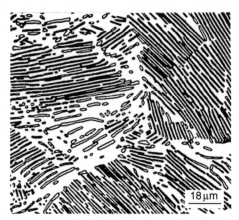

Figure 14.9 Fully pearlitic structure showing dark-etching layers as iron carbide and the light-etching layers as ferrite. From W. D. Callister, Jr. and D. G. Rethwisch, *Fundamentals of Materials Science and Engineering*, John Wiley & Sons, New York, 2008, 385. Reprinted with permission of John Wiley & Sons, Inc. Copyright held by John Wiley & Sons, Inc.

The leftmost arrow in Figure 14.8 notes that, upon slow cooling, austenite will transform to a structure involving a two-phase mixture of ferrite and cementite. These two phases are juxtaposed in a layered structure called pearlite. Pearlite consists of two phases, and the combination is properly called a "microconstituent." The characteristic pearlitic micro-structure is shown in Figure 14.9.[102] Now, unless the steel has a composi-tion of 0.80% carbon (the eutectoid composition), some bulk ferrite or carbide will appear along with the pearlite. This additional material is referred to as *proeutectoid*, since it forms above the eutectoid temperature. Figure 14.10 shows a microstructure containing a combination of pro-eutectoid ferrite and pearlite.[103] Steels that are heated into the austenite region and slowly cooled to form a proeutectoid phase and pearlite are said to have been *normalized*.

The center arrow in Figure 14.8 indicates that upon moderate cooling the austenite will transform to a ferrite plus cementite microconstituent called *bainite*. Bainite involves a very fine dispersion of cementite particles in a fine-grained "feathery" appearing ferrite structure. If the austenite is cooled rapidly enough, or quenched, the austenite will not form a combination of ferrite and cementite, but will dynamically transform to a body-centered structure with carbon atoms trapped similar to solute atoms. This phase is called *martensite*, and is, in effect, ferrite supersaturated with carbon. Mar-tensite is unstable, and does not appear on the phase diagram in Figure 14.4.

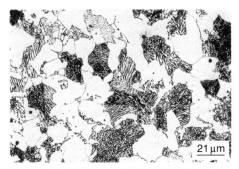

Figure 14.10 Microstructure with comparable amounts of pearlite and ferrite with ferrite as the light continuous phase. From W. D. Callister, Jr. and D. G. Rethwisch, *Fundamentals of Materials Science and Engineering*, John Wiley & Sons, New York, 2008, 387. Reprinted with permission of John Wiley & Sons, Inc. Copyright held by John Wiley & Sons, Inc.

14.4.2 Tempered martensite

Martensite, although hard, is quite brittle and impractical for most uses. However, its toughness can be greatly improved by *tempering*. This process involves heating the martensite to a temperature in the 250 to 650°C range. In this process, represented at the lower right in Figure 14.8, fine carbides precipitate from the martensite, leading to a ferrite-cementite structure similar to that of bainite. Figure 14.11 displays the effects of tempering temperature on the mechanical properties of tempered martensite.[104]

14.4.3 Spheroidite

Actually, none of the austenite transformation products shown in Figure 14.8 is stable, and if the transformation products are held for many hours at a temperature just below 732°C, the cementite will emerge as a globular or spherical phase in a ferrite matrix. This structure is called *spheroidite*, and a typical microstructure is shown in Figure 14.12.

Thermal treatment below the eutectoid temperature of 732°C is generally referred to as an *anneal*.

14.5. STRUCTURE-MECHANICAL PROPERTY RELATIONS

All else equal, increased carbon content increases the strength of steel. However, the microstructure plays an important role as well. These points are illustrated in Figure 14.13.[105] A general listing of typical tensile strengths for various steel microstructures and process histories is given in Table 14.1.

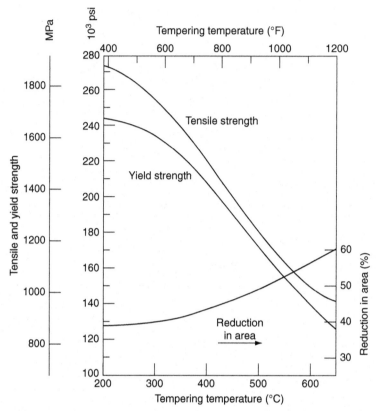

Figure 14.11 The effects of tempering on the mechanical properties of martensite. From W. D. Callister, Jr. and D. G. Rethwisch, *Fundamentals of Materials Science and Engineering*, John Wiley & Sons, New York, 2008, 436. Reprinted with permission of John Wiley & Sons, Inc. Copyright held by John Wiley & Sons, Inc.

 ## 14.6. TRANSFORMATION DIAGRAMS

14.6.1 Isothermal transformations

The transformation of austenite to pearlite does not happen instantaneously, and the time required varies depending on the point of cooling below the 732° C transformation temperature. This behavior is quantified in Figure 14.14.[106] Figure 14.14 is called an *isothermal transformation diagram,* and it represents the behavior of a eutectoid steel, or a steel with 0.80% carbon. As an example, suppose that the steel is very rapidly cooled from 800 to 600°C. It then goes from a temperature at which austenite is stable to a temperature at which austenite still exists, but is unstable. The region of austenite existence in the

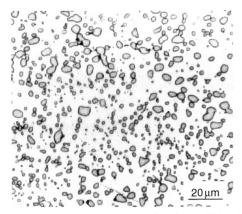

20 µm

Figure 14.12 Spheroidized microstructure with globular carbides in a ferrite matrix. From W. D. Callister, Jr. and D. G. Rethwisch, *Fundamentals of Materials Science and Engineering*, John Wiley & Sons, New York, 2008, 420. Reprinted with permission of John Wiley & Sons, Inc. Copyright held by John Wiley & Sons, Inc.

transformation diagram is labeled A. Now, if the steel is held isothermally at 550°C, it will start to transform to pearlite in a little less than one second, and the transformation to pearlite will be complete in about seven seconds.

The transformation illustrated in Figure 14.14 is very important because it creates a very-fine-grained pearlite structure, which possesses excellent drawability, en route to very high strength. It is the basis for the manufacture of tire cord and other ultra-high-strength steel wire products. Isothermal processing to achieve this very-fine-grained pearlite is called *patenting*. Historically, the isothermal condition has been achieved by quenching into molten lead, held at the desired transformation temperature.

If the transformation occurred at 650°C, a coarser pearlite structure would have developed with lower strength potential and less drawability. An isothermal transformation at, about 400°C would have produced bainite.

In general, higher carbon steels have a structure that is primarily pearlite under drawing conditions, and low carbon steels are drawn with a structure that is primarily ferrite, consistent with the vertical lines in Figure 14.4. All structures become stronger with increased drawing reduction, and an unworked structure can be restored with annealing or normalizing.

14.6.2 Continuous cooling (non-isothermal) transformations

While isothermal processing is certainly industrially feasible and allows for rather precise process control, it is often expeditious to use non-isothermal, continuous cooling transformation instead. Figure 14.15 displays the

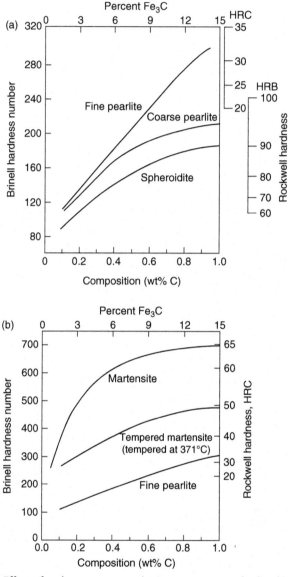

Figure 14.13 Effect of carbon content and microstructure on the hardness of steel. From W. D. Callister, Jr. and D. G. Rethwisch, *Fundamentals of Materials Science and Engineering*, John Wiley & Sons, New York, 2008, 432–433. Reprinted with permission of John Wiley & Sons, Inc. Copyright held by John Wiley & Sons, Inc.

Table 14.1 Typical mechanical properties for a variety of steel compositions and structures

Composition, Microstructure, Process History	Tensile Strength
Very low C (0.06%), ferrite, normalized	360–405 MPa
Medium C (0.30%), ferrite/pearlite, normalized	640–720
Medium C (0.30%), ferrite/pearlite, cold drawn	1000–1500
High C (0.7%), pearlite, normalized	1050–1125
High C (0.7%), pearlite, cold drawn	2000–3000
Medium C (0.40%), bainite	1500–2000
Medium C (0.40%), tempered martensite	1000–2000
Medium C (0.40%), spheroidite	400–500

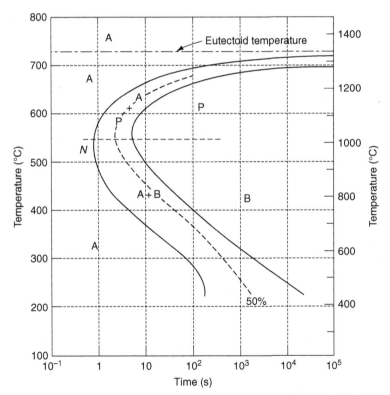

Figure 14.14 Isothermal transformation diagram for a eutectoid steel (0.76% carbon). From W. D. Callister, Jr. and D. G. Rethwisch, *Fundamentals of Materials Science and Engineering*, John Wiley & Sons, New York, 2008, 420. Reprinted with permission of John Wiley & Sons, Inc. Copyright held by John Wiley & Sons, Inc.

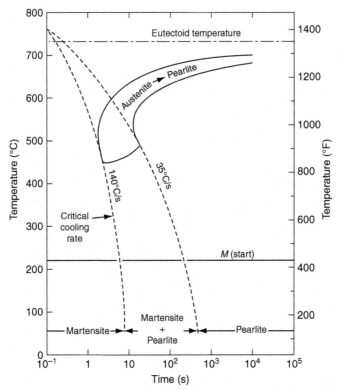

Figure 14.15 Transformation diagram for continuous cooling responses for a eutectoid steel (0.76% carbon). From W. D. Callister, Jr. and D. G. Rethwisch, *Fundamentals of Materials Science and Engineering*, John Wiley & Sons, New York, 2008, 429. Reprinted with permission of John Wiley & Sons, Inc. Copyright held by John Wiley & Sons, Inc.

transformation responses for continuous cooling for a eutectoid steel.[107] The Stelmor® process uses such an approach, developing pearlitic microstructures for increased drawability or increased descaling response.[108]

14.6.3 Forming martensite

Cooling more rapidly than the critical cooling rate of 140°C indicated in Figure 14.15 involves cooling rates that avoid the austenite–pearlite transformation region. Thus, the pearlite transformation is entirely suppressed. Once the temperature reaches the M_s temperature at approximately 220°C, the austenite will start to transform, nearly instantaneously, into martensite. This transformation is not time dependent, but it is temperature dependent; that is, as the temperature gets lower, more martensite forms. Eventually, at a temperature referred to as M_f, the transformation to martensite should be complete. The level of the M_f may be somewhat obscure; however, and martensite transformation temperatures are often identified with the degree of martensite transformation. For example, M_{90} would indicate the temperature at which 90% of the martensitic transformation has occurred (roughly 100°C for a eutectoid steel).

Alloying has a substantial effect on the curves in Figure 14.15. In nearly all cases, alloying lowers the martensite transformation temperatures. Alloying also delays the transformation to pearlite, or displaces the "noses" or the "C-curves," representing pearlite transformation, to higher times. This latter effect can be very useful in the thermal processing of steels, since slower, more easily achieved cooling rates can result in avoidance of pearlite, and full transformation to martensite can occur. Steels with slower pearlite transformation and with C-curve noses displaced to longer times are said to be more *hardenable*, and the alloying additions that accomplish this end are said to increase *hardenability*.

The effects of alloying elements in lowering the martensite transformation temperature and in delaying pearlite transformation do not occur for cobalt. Moreover, at higher levels increased carbon will not improve hardenability.

 ## 14.7. FLOW STRESS, COLD WORKING, AND ANNEALING

14.7.1 The flow stress of steel

The strength of steel, as previously discussed, varies enormously with composition and microstructure. Nonetheless, for cold working purposes, annealed and normalized steel strength data can, as with other metals, be fitted to and projected from the equation $\sigma_o = k\varepsilon_o^N$. This will generally require laboratory testing dedicated to the properties of the steel of interest.

Consider the following examples. Figure 14.16 displays an extraordinary range of flow stress versus strain relations for very low carbon iron (0.007% C).[109] The stress–strain relations can be approximated by the relation:

$$\sigma_o(MPa) = 430\varepsilon_o^{0.59}. \qquad (14.7)$$

The stress–strain relationship for 1020 steel (0.20% C) in Figure 11.8 can be approximated by

$$\sigma_o(MPa) = 725\varepsilon_o^{0.40}. \qquad (14.8)$$

The stress–strain relationship for 0.27% C steel in Figure 14.17[37] can be approximated by

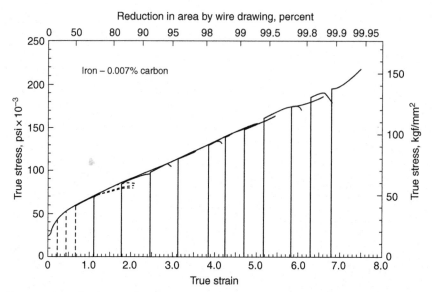

Figure 14.16 Flow stress for 0.007% carbon iron. From G. Lankford and M. Cohen, *Transactions of the American Society for Metals*, 62 (1969) 623. Copyright held by ASM International, Materials Park, OH.

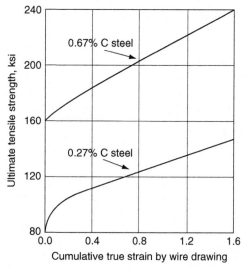

Figure 14.17 Tensile strength versus true strain curves for 0.27 and 0.67% carbon steel wires, drawn with successive 20% reductions. Data taken from A. B. Dove, Deformation in Cold Drawing and Its Effects, *Steel Wire Handbook*, Vol. 2, The Wire Association, Inc., Branford, CT, 1969, 16.

$$\sigma_o(\text{MPa}) = 960\varepsilon_o^{0.23}. \tag{14.9}$$

Finally, the stress-strain relationship for 0.67% C steel in Figure 14.17 can be approximated by

$$\sigma_o(\text{MPa}) = 1485\varepsilon_o^{0.19}. \tag{14.10}$$

There is a trend in these data-fitting equations for N to decrease and for k to increase with increased carbon content. However, such generalizations should not be a substitute for running a few tests on the material of interest. The approach illustrated in Figures 14.16 and 13.3 (pulling tensile tests after each drawing pass and superimposing the curves to make a master curve) is the recommended method.

14.7.2 Annealing cold worked steel

Understanding the annealing procedures for cold worked steel requires another look at the iron-iron carbide phase diagram in Figure 14.4. Annealing with the simple objective or removing cold work is generally called *process annealing*. It is undertaken *below* the eutectoid temperature (732°C) so austenite and its transformation products are not inadvertently formed. However, temperatures as high as 700°C can be used. A nominal time of one hour, at temperature, is typical of such anneals. Extended time anneals just below the eutectoid temperature are used to establish spheroidized structure, as noted in Section 14.4.3.

Where relief of residual stresses is desired, without necessarily recrystallizing the structure, anneals at 550°C or lower are used. Again a nominal time of one hour at temperature is typical.

It was noted in Section 14.4.1 that heating of steel into the austenite region, followed by relatively slow cooling, is called *normalizing*. Another thermal treatment that should be noted, involving heating above the eutectoid temperature, is the *full anneal*. This process involves heating at about 50°C above the temperature at which the steel becomes nearly all austenite, and then cooling very slowly ("furnace cooling"). This procedure forms relatively soft, coarse pearlite, which may be advantageous for certain machining or forming operations.

14.8. AGING IN STEEL

The behavior discussed previously is often complicated by the relocation of carbon (and nitrogen) atoms over time at ambient temperature, and over very short times at moderately elevated temperature. This topic was first mentioned in Section 11.2.5, and Figure 11.7 describes the yield point phenomenon, which reflects the role of *aging* in steel. A comprehensive discussion of this subject follows. The author has published a practical, wire-industry-focused review of aging in steel.[110]

14.8.1 Introduction

As a metallurgical term, aging usually refers to an increase in strength and hardness, coupled with a decrease in ductility and toughness that occurs over time, at a given temperature. In some cases these changes achieve a maximum or minimum at a certain time with conditions at later times referred to as "overaged."

Such aging is usually attributable to changes in the state or location of solute elements. There are two major modes: aging that involves precipitation from a supersaturated solid solution is referred to as *quench aging*, and *strain aging* involves the redistribution of solute elements to the strain fields of dislocations. When strain aging occurs in the absence of concurrent plastic deformation it is called *static*, and when strain aging occurs concurrently with plastic deformation it is called *dynamic*. Generally, the rate of aging increases with temperature, unless the solubility of the solute element changes substantially. In the temperature range of dynamic strain aging, the aging response is quite rapid. It should be noted that quench aging and strain aging may coexist.

In steel, the practical aging responses involve the role of carbon and nitrogen in ferrite (or in martensite, if tempering is viewed as an aging process). There are many nuances of this type of aging in steel wire processing and related mechanical properties. Strain aging, in particular, is pertinent to this chapter.

14.8.2 Static stain aging in steel

Strain aging in steel involves the positioning of carbon and nitrogen atoms that are actually in solution in ferrite. As shown in Table 14.2, the population of these atoms should be quite low at ambient temperature, especially for the case of carbon.[111] Nonetheless, the phenomenon of strain aging is marked, even at ambient temperature.

Basically, any interstitial solute carbon or nitrogen atom exerts a certain *compression* on the surrounding crystal structure. Moreover, dislocation stress fields involve regions of *tension*. Thus, the mechanical energy of the steel can be lowered if carbon and nitrogen atoms migrate into the regions of tension associated with the dislocations. Since energy is

Table 14.2 Approximate temperatures for 0.01, 0.001, and 0.0001% solubility limits of iron carbide and nitrides in ferrite

Compound	0.010 (°C)	0.001 (°C)	0.0001 (°C)
Fe_3C	600	450	300
Fe_4N	350	200	100
$Fe_{16}N_2$	200	100	—

From *The Making, Shaping and Treating of Steel*, 10th Edition, W. T. Lankford, Jr., N. T. Samways, R. F. Craven and H. E. McGannon (Eds.), Association of Iron and Steel Engineers, Pittsburgh, PA, 1985, 1284. With permission.

lowered, this migration occurs naturally with time. However, the dislocations must now increase the mechanical energy of the steel if they are to move away from the preferentially located carbon and nitrogen atoms. This requires extra stress for dislocation motion, and means that the steel has a higher yield strength. The increase in yield strength associated with strain aging is manifested in an upper yield point, as shown in Figure 14.18 at about 240 MPa.[112] The data in Figure 14.18 are from a "box-annealed" steel, where the slow cooling ensures a pronounced strain aging. As the tensile stress reaches the upper yield point, dislocations

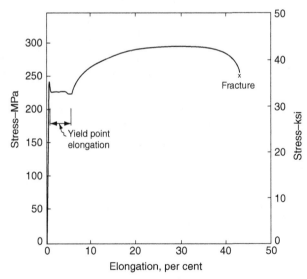

Figure 14.18 Stress-strain curve for "box annealed" carbon steel. *The Making, Shaping and Treating of Steel*, 10th Edition, W. T. Lankford, Jr., N. T. Samways, R. F. Craven and H. E. McGannon (Eds.), Association of Iron and Steel Engineers, Pittsburgh, PA, 1985, 1400. Copyright held by Association for Iron & Steel Technology, Warrendale, PA, USA.

presumably are able to move away from their "atmospheres" of carbon and nitrogen. The moving dislocations multiply, as well, and a large population of unencumbered dislocations allows deformation to occur more readily, resulting in a drop in stress to the lower yield point shown in Figure 14.18 at about 225 MPa. This shift occurs in stages, involving Lüders bands in the steel, as noted in Figure 11.7b. Eventually the yield point elongation is achieved and normal strain hardening ensues. The magnitude of the upper yield point depends on the stiffness of the testing machine and the strain rate of the tensile test.

The yield point phenomenon can be a major nuisance in many forming operations (resulting in Lüders lines or stretcher strains, "fluting," discontinuous yielding, etc.). The yield point can be removed by deforming the steel to just beyond the yield point elongation. This is especially common with sheet and strip processing for deep drawing purposes, for which a light "temper" rolling pass, or "skin pass", is used, as illustrated in Figure 14.19. In wire processing, a light sizing pass or a straightening pass may accomplish the same objective. It is essential to remember, however, that strain aging will not stop, and within a relatively short time the yield point will be observed again; that is, there will be a time window for any yield-point-free forming to be accomplished.

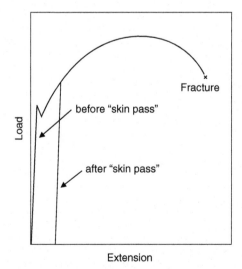

Figure 14.19 Illustration of the change in load-elongation curve with a light "skin pass."

Table 14.3 Approximate times for full strain aging in carbon steels at different temperatures

Temperature (°C)	Time
0°C	20 wks
21	10 wks
100	1½ hrs
120	20 min
150	4 min

The time for strain aging is a strong function of temperature. The degree of strain aging represented in Figure 14.18 can be achieved in about 1½ hours at 100°C, and Table 14.3 provides some time-temperature relationships for substantial aging. In this context, it is important to note that strain aging does not continue into an overaged mode, and, as full strain age strengthening is achieved, no subsequent decrease in strength can be expected with time.

14.8.3 Dynamic strain aging in steel

Extrapolation of the data of Table 14.3 to higher temperatures indicates that times can be so short they involve strain aging concurrent with plastic deformation, even at practical processing and tensile testing strain rates. When strain aging occurs concurrently with plastic deformation it is called dynamic.

Dynamic strain aging in steel, at processing strain rates, occurs most noticeably in the 200 to 650°C range, as displayed in Figure 14.20.[113] The flow stress or strength data for a strain rate of $6.6 \times 10^{-2}s^{-1}$ shows a marked peak at about 270°C, and the strength data for strain rates of $10s^{-1}$ and $430s^{-1}$ show peaks at about 400°C and 500°C, respectively. These peaks result from the inhibition of dislocation motion by strain aging from carbon and nitrogen atoms, and the carbon and nitrogen are so mobile that moving dislocations cannot pull away. The decline in strength with temperature above the peaks in Figure 14.20 may reflect the onset of recovery mechanisms.

The impact of strain rate on the peak positions in Figure 14.20 creates a region of *anomalous strain rate effect* in the 250°C range; that is, in most cases increased strain rate leads to increased strength, but in this region it is the opposite. In the anomalous strain rate region increased strain rate produces faster average velocities for dislocations and/or increased numbers of

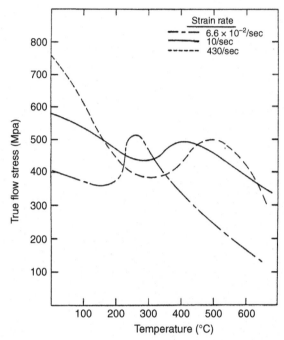

Figure 14.20 Stress-strain rate-temperature relations for a low carbon steel displaying the anomalous strain rate effect due to dynamic aging. From M. E. Donnelly, The Flow Stress of Low Carbon Steels Under Hot Deformation: A Basic Data Compilation for Process Analysis, M. S. Project, Rensselaer Polytechnic Institute, 1982.

dislocations in such a way that the strain aging effects of carbon and nitrogen cannot keep up. Hence the steel has *decreased strength*, not increased strength, at the higher strain rate.

14.8.4 Relevance to Pearlite-Containing Steels

The previous discussions have focused on the state of carbon and nitrogen in ferrite alone, and it is in ferrite that the aging processes occur. The dominance of such behavior in very low carbon steel or spheroidized steel is obvious. However, ferrite is also critical to the mechanical behavior of pearlite-containing steels, and such steels display the aging responses. This is not surprising, since fully pearlitic, eutectoid steels are nearly 90% ferrite.

14.8.5 Implications for wire processing

First, all carbon steel wire conditions are prone to static strain aging, to the consequent appearance of yield point phenomena, fluting (flattening in bending), and discontinuous yielding in forming operations. Moreover, any carbon steel wire that has been in service more than a few months at ambient temperature may be assumed to be fully strain aged.

Dynamic strain aging is a major consideration in steel wire processing, since thermomechanical heating and limited cooling capability can easily result in deformation temperatures of 200°C and higher. Moreover, such temperatures are apt to be unstable and result in varying degrees of dynamic strain aging. Much effort has been devoted to this, including highly engineered cooling systems, modeling and measurement of average temperatures during drawing, tapered drafting schedules, and so forth. It is important to note how ASTM Designation A 648-04a, "Steel Wire Hard Drawn for Prestressing Concrete Pipe," addresses the matter of wire temperature and strain aging. Subsection 4.3 of Designation A 648-04a is quoted below.

4.3 The wire shall be cold drawn to produce the desired mechanical properties. The wire manufacturer shall take dependable precautions during wire drawing to preclude detrimental strain aging of the wire.

Note 2 — Allowing wire to remain at elevated temperatures, such as 400°F (204°C) for more than 5 s or 360°F (182°C) for more than 20 s, can result in detrimental strain aging of the wire. Detrimentally strain aged wire typically has reduced ductility and increased susceptibility to hydrogen embrittlement.

Designation A 648-04a refers to strain aging in general, and lists times at temperature that would be relevant to *static* strain aging. However, the implications are that wire drawing is occurring at higher temperatures than temperatures that would be pertinent to 5 or 20 s duration.

The temperatures for dynamic strain aging in steel wire drawing are dependent on strain rate. The strain rate in wire drawing is discussed in Section 7.2.2. The strain rate range of 10^3 to 10^4 s^{-1} corresponds with commercial steel drawing practice. On the basis of the data in Figure 14.20, the indicated dynamic strain aging range would be about 500 to 650°C. While this temperature range is exceptionally high for the average drawing temperature in sophisticated steel drawing practice, it is not unreasonable for the wire surface temperature, as discussed in Section 6.1.7. Such temperatures only exist for a short time, but time is not a variable in the case of dynamic strain aging.

The ASTM Designation A 648-04a guidelines have emerged from carefully developed empirical databases, and are clearly a basis for careful practice. They likely minimize immediate static strain aging, and may require that dynamic strain aging conditions be avoided for the bulk of the wire cross section during drawing. However, it seems that the surface of the wire may have been dynamically strain aged in any case.

In analysis of practical situations, the dynamic strain aging of the wire surface must be considered in conjunction with such issues as lubricant breakdown, crow's feet development, oxidation, and even austenitization (and subsequent martensite development). Compromised wire properties from excessive drawing temperature cannot, in these contexts, be attributed solely to strain aging.

It is especially important, however, to note that lower strain rate drawing, forming, fastener manufacturing, and so forth, need not reach the 500 to 650°C range for dynamic strain aging to occur. Indeed, temperatures in the 200 to 400°C range will allow dynamic strain aging conditions to occur in operations of modest deformation rate (Figure 14.20).

Strain aging is actually usefully promoted by so-called "low temperature" stress-relieving heat treatments. Technically, a ferrous stress-relieving heat treatment should be a thermal exposure below the austenite range that relieves internal stresses by allowing dislocation motion over time. The dislocations are driven by the internal stresses and move with time, producing creep. The creep strains relieve the driving stresses, and in time the stresses are substantially relaxed. Classical carbon steel stress relieving has involved such treatments as one hour at 540°C, and so forth. In the steel wire industry, it has been common practice to use the term stress relieving for heat treatments as low as 200°C. Such heat treatments maximize strength and minimize stress relaxation in cold drawn wire. However, it should be pointed out that this property improvement almost certainly results from the rapid static strain aging that occurs with these heat treatments. It is difficult to believe that much dislocation motion occurs at such a low temperature, especially in the context of dislocation immobilization by way of static strain aging. In any event, the low temperature stress relief heat treatment can be beneficial where stress relaxation is a problem in fasteners, cables, reinforcing wires, and other products that bear high sustained stresses.

14.8.6 Control of aging

Development of aging in the short term can be minimized by avoiding some of the contributing processing conditions noted earlier. The introduction of a light temper or straightening pass will temporarily remove the effects of aging, as illustrated in Figure 14.19. Strain aging will soon return, however, as shown in Table 14.3.

From a compositional or alloy perspective, it is impractical to reduce aging simply by lowering carbon and nitrogen. However, aging can be greatly reduced by tying up nitrogen and carbon as stable nitrides and carbides. For example, as little as 0.05 w/o aluminum can, with proper processing, tie up nitrogen as AlN, and titanium and niobium additions have long been used to tie up carbon.

 ## 14.9. CARBON STEEL COMPOSITIONS

A generally useful system of steel designation is the Unified Numbering System (UNS), which incorporates the traditional Society of Automotive Engineers (SAE) and American Iron and Steel Institute (AISI) systems. For carbon steels, the UNS designation starts with the letter "G," followed by the four numbers of the SAE-AISI system, usually followed by 0. There are many international steel designations, as well, but they can usually be correlated with the UNS system. To simplify notation, this text uses the four numbers of the SAE-AISI system, or 10XX for most carbon steels, where the XX designation indicates the weight percent carbon in hundredths of a percent. For example, plain carbon steel with 0.20 weight percent carbon content would have an SAE-AISI designation of 1020 and a UNS designation of G10200.

Plain carbon steel chemical specifications usually call out only four elements (other than iron): carbon, manganese, sulfur, and phosphorous. Carbon increases strength and, up to intermediate levels, increases hardenability. Manganese also increases strength and hardenability, and is beneficial to surface quality in most cases. Increased carbon and manganese can compromise ductility and weldability, however. Phosphorous and sulfur generally decrease ductility and toughness, an especially big problem in the early days of steel making, and their chemical specifications involve maximum tolerances in the case of 10XX plain carbon steels. As an example, the specification for SAE-AISI 1020 or UNS G10200 steel is as follows (with compositions in weight per cent): 0.18– 0.23% carbon, 0.30–0.60% manganese, 0.040% maximum phosphorous, and 0.050% maximum sulfur.

There are three other carbon steel specification categories, 11XX, 12XX, and 15XX, using the SAE-AISI system. The 15XX system involves increased manganese additions. The 11XX "free cutting" or re-sulfurized system involves sulfur typically in the range of 0.10%, where the sulfur greatly improves machinability through the effect of finely dispersed sulfides. Finally, phosphorous also improves machinability, and the 13XX system involves both sulfur and phosphorous, and sometimes lead, additions for that purpose.

It is important to understand that carbon steels contain numerous additional "residual" elements in their composition. In general these elements are not specified, even when it is understood that their levels are not small. Of significant interest is silicon, which is retained mostly from its role in steelmaking deoxidation practice. Levels of residual silicon can range from negligible to as much as 0.60%. Other important residual elements can be traced to pig iron and, especially, steel scrap used in primary processing of steel. Such elements include chromium, nickel, copper, molybdenum, aluminum, vanadium, and so on. The sum total of the residual elements can be the order of 1.00% in carbon steels. It is generally assumed that these elements are "harmless" or even "beneficial." Certainly they generally increase strength. However, their impact on ductility, toughness, weldability, and surface quality may be problematical.

14.10. LOW ALLOY STEEL COMPOSITIONS

The majority of desirable steel property combinations can be developed with plain carbon steels, provided there is adequate processing capability or limited product dimensions. However, plain carbon steels that are to be hardened (by martensite formation) require relatively rapid cooling, which may not be practical for the interior of a larger cross section. Moreover, such rapid cooling may involve the risk of distortion or cracking. Alloying generally improves hardenability, or allows slower cooling to be undertaken en route to the formation of martensite. A great improvement in hardenability can be achieved with minimal alloying (a very few percent of alloying additions), and many low alloy steels or simply alloy steels, have been designed with this in mind. There are numerous categories of lightly alloyed steels addressing objectives such as corrosion resistance and electromagnetic response; however, this discussion will focus on alloying for hardenability and general mechanical property improvement.

Manganese is generally present in low alloy steels, as it is in plain carbon steels, and its role in improving hardenability has been cited. Chromium is perhaps the most prominent alloying addition to low alloy steels. Chromium greatly improves hardenability, and its carbides are hard and stable at higher temperatures, thus expanding the temperature range for the use of the steel. The SAE-AISI designation for chromium low alloy steels is 50XX or 51XX, where the XX designation indicates the carbon content, in hundredths of a percent.

Molybdenum is highly effective in increasing the hardenability of low alloy steels. Moreover, it largely eliminates the reduced toughness ("temper embrittlement") often observed in low alloy steels upon tempering exposures in the 250–400°C range. The SAE-AISI designation for molybdenum steels is 40XX. Chromium and molybdenum additions are often used together ("chrome-moly" steels), and the relevant SAE-AISI designation is 41XX.

Nickel additions generally add to the robustness of low alloy steels in that nickel inhibits some of the grain coarsening tendencies associated with chromium additions, and generally leads to increased toughness, or a lowering of the ductile-to-brittle transition temperature. Chromium-nickel-molybdenum low alloy steels are preferred for many applications, and have the relevant SAE-AISI designations of 43XX and 86XX, among others.

Other alloying elements are often employed in improving the processing latitude and properties of hardenable low alloy steels, particularly vanadium and boron. In addition to increasing hardenability, vanadium refines grain size and forms strong, stable carbides. The relevant SAE-AISI designation for chromium-vanadium low alloy steels is 61XX. Boron can be extremely potent in promoting hardenability, with a common range of 0.0005–0.003 weight percent. Boron steels are generally designated by placing a B before the number indicating carbon content, such as 86B45 indicating a chromium-nickel-molybdenum low alloy steel containing a boron addition and about 0.45% carbon.

14.11. QUESTIONS AND PROBLEMS

14.11.1 Figure 14.11 shows the effects of tempering on the mechanical properties of an alloy steel. Can this steel be tempered to manifest at least 45% area reduction at fracture with a tensile strength minimum of 1150 MPa?
Answer: Yes, tempering below 525°C will ensure a tensile strength of at least 1150 MPa, and tempering above 440°C will ensure an area reduction of at least 45%. Thus, a tempering treatment between 440 and 525°C should work.

14.11.2 Consider Figure 14.14 and an isothermal treatment at 650°C. When will pearlite start to form, and when will pearlite formation be complete?

Answer: Placing a line at 650°C, and recognizing that the time scale is logarithmic, it appears that pearlite will start to form at about 8 seconds and finish forming at about 80 seconds.

14.11.3 What appears to be the dependence of uniform elongation on carbon content in carbon steels?

Answer: One approach to answering this question is to realize that when stress–strain relations are expressed as $\sigma_o = k\varepsilon_o^N$, it can be shown that the uniform elongation actually equals N (note Equation 11.11). Now looking at Equations 14.7 through 14.10, it is clear that respective carbon levels of 0.007%, 0.20%, 0.27%, and 0.67% correspond to respective uniform elongation levels of 0.59, 0.40, 0.23, and 0.19. Thus, uniform elongation decreases as carbon content increases.

14.11.4 Table 14.2 displays three solubility limits for Fe_3C in iron. Using the phase diagram of Figure 14.4, add a fourth solubility limit.

Answer: Near the lower left-hand corner of the diagram it is clear that a carbon content solubility limit of 0.035% is associated with ferrite at 732°C.

CHAPTER FIFTEEN

Other Metallurgical Systems for Wire Technology

Contents

15.1. ALUMINUM AND ITS ALLOYS

15.1.1 Important properties of aluminum

Aluminum has low density, as seen in Table 6.1. Thus its properties, on a per-weight basis, can be quite competitive. An important example is conductivity. The conductivity of aluminum electrical conductor can be over 60% of that of copper, and yet the density of aluminum is only about 30% that of copper. Thus, the conductivity-to-weight ratio for aluminum can be over twice that of copper. The density of aluminum is only about 35% that

Wire Technology
ISBN 978-0-12-382092-1, DOI: 10.1016/B978-0-12-382092-1.00015-4

of steel, making aluminum fully comparable to steel on a stiffness-to-weight basis, and competitive with steels nearly three times stronger on a strength-to-weight basis.

Aluminum has great resistance to corrosion in natural atmospheric environments, in spite of its high chemical activity. This arises from the tightly adhering, relatively impervious oxide coating that forms on the aluminum surface.

Aluminum is a face-centered-cubic (FCC) metal with great workability.

15.1.2 Primary processing

Aluminum is quite plentiful in the earth's surface, but generally present in oxygen-containing chemical forms; as a result, it is difficult to separate it from its ores. Its widespread availability as an engineering metal did not develop until the early years of the twentieth century.

As an initial step, aluminum oxide, or bauxite, is reacted with hot sodium hydroxide to form sodium aluminate, and after separation from ore residue, the aluminate is slowly cooled. The nominal equations are

$$Al_2O_3 + 2NaOH \rightarrow 2NaAlO_2 + H_2O$$
$$NaAlO_2 + 2H_2O \rightarrow Al(OH)_3 + NaOH \tag{15.1}$$

The resultant aluminum hydroxide is calcined at 1100°C to produce bulk aluminum oxide.

The aluminum oxide is next dissolved in molten cryolite (Na_3AlF_6) and electrolyzed using carbon anodes and cathodes. Molten aluminum at 99.5 to 99.9% purity can be produced. The general process is called the Bayer process, and the electrolysis is called the Hall-Heroult process. The product of this process is generally re-melted and processed into rod using techniques similar to those outlined in Chapter 13 for copper. A detailed discussion of redraw rod processing is presented in Chapter 17.

Aluminum primary processing is quite energy intensive, and generally located where there is a readily available and/or economic energy source. Scrap and recycling practice play major roles in aluminum economics.

15.1.3 Flow stress, cold work, and annealing

Stress-strain curves for a number of aluminum alloys and conditions are illustrated in Figure 11.8. Using the relationship $\sigma_o = k\varepsilon_o^N$, the data for three as-annealed cases and one cold worked case are listed in Table 15.1.

Table 15.1 Flow curve representations for three as-annealed aluminum alloys and one cold worked aluminum alloy

Composition	Condition	k value (MPa)	N value
1100 Al: 99.00% Al, 0.12% Cu	Annealed	95	0.41
1100-H14 Al: 99.00% Al, 0.12%Cu	Cold worked	240	0.29
6061 Al: Al-Mg-Si-Cu-Cr alloy	Annealed	235	0.22
2024 Al: Al-Cu-Mg-Mn alloy	Annealed	435	0.12

Low alloy content aluminum is generally annealed at about 345°C to remove the effects of cold work. Annealing temperatures for a number of common aluminum alloys are about 415°C.

In addition to the obvious dependence on strain, aluminum flow stress is also dependent on temperature and strain rate. The effect of strain rate, $d\varepsilon_o/dt$, based on various data in the literature, is reasonably expressed by Equation 15.2:

$$\sigma_o(MPa) = G(d\varepsilon_o/dt)^{0.0155}, \qquad (15.2)$$

where G is the strain rate strength coefficient. The effect of temperature is reasonably expressed by Equation 15.3:

$$\sigma_o(MPa) = \sigma_{To} + (-0.29)(T-T_o), \qquad (15.3)$$

where σ_{To} is the flow stress in MPa at the reference temperature T_o, in °C. Figure 15.1 displays some relationships among flow stress, strain rate, and temperature for aluminum.[68]

15.1.4 Crystal structure and phase relations

As previously noted, aluminum has an FCC structure, and its great workability reflects this. Aluminum intended for electrical applications generally has very low alloy content. Some aluminum alloys are alloyed for strength utilizing solid solution strengthening mechanisms. Other aluminum alloy systems are designed for precipitation or age hardening.

The FCC phase is dominant in all strengthening mechanisms. With solid solution strengthening, the solute element simply dissolves into the solvent FCC phase. In the case of age hardening alloys, the alloying elements are solution treated, or dissolved into the FCC phase at an elevated temperature, followed by precipitation as compounds from the FCC phase at a lower temperature. In *natural aging*, precipitation is allowed to occur at room

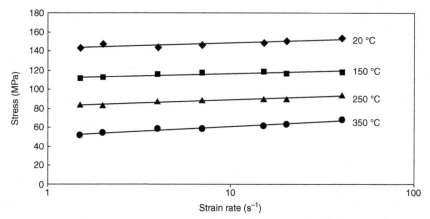

Figure 15.1 Relationships among flow stress, strain rate, and temperature for aluminum. From J. F. Alder and V. A. Phillips, *Journal of the Institute of Metals*, 83 (1954–1955), 82. Copyright held by Maney Publishing, London, UK, www.maney.co.uk/.

temperature. In other cases, the alloy is designed for an elevated temperature treatment, and this process is called *artificial aging*.

15.1.5 Aluminum and aluminum alloy grades and temper designations

Wrought aluminum and aluminum alloy grades are described by the Aluminum Association system of four digits, such as 1100. In general, UNS designations involve A9 followed by the Aluminum Association number, such as A91100. The Aluminum Association designation system is outlined in Table 15.2.

Aluminum products are generally accompanied by a *temper designation*; the word temper refers to a process condition, and is not to be confused with the use of the term temper as used in processing of martensitic

Table 15.2 Basis for Aluminum Association designations of wrought aluminum alloys

Designation Number	Principal Alloying Elements
1xxx	None, 99% + aluminum
2xxx	Copper
3xxx	Manganese
4xxx	Silicon
5xxx	Magnesium
6xxx	Magnesium and silicon
7xxx	Zinc
8xxx	Other

Table 15.3 Basic temper designations for aluminum and its alloys

Temper Designation	Process Condition
F	As-fabricated
O	Annealed
H	Strain hardened
W	Solution heat treated, with natural aging
T	Solution heat treated, for artificial aging

steel. The basic temper designations for aluminum and its alloys are listed in Table 15.3. The designations listed generally involve additional numbers, designating process details or subcategories.

15.1.6 Aluminum and aluminum alloy grades important to wire technology

A large component of aluminum wire production is for electrical applications. The classical grade for electrical applications is 1350, or "EC aluminum." The aluminum content is at least 99.5%, and the conductivity is at least 61.8% IACS in the annealed condition.

A number of aluminum grades used in wire form are summarized in Table 15.4.

Table 15.4 Aluminum grades used in wire form

Designation Number	Principal Alloying Elements	Applications
1050, 1060, 1100	None	General and miscellaneous, including rivet wire, weld wire, spray gun wire, electrical
1350	None	Electrical
2011	Cu, Pb, Bi	Screw machine products
2024	Cu, Mg, Mn	Rivets, screw machine products
2319	Cu, Mn, Zr, Ti, V	Electrodes and filler wire
4043	Si	Weld filler wire
5005	Mg	Electrical
5052	Mg, Cr	Rivets
5056	Mg, Mn, Cr	Screens, nails, rivets
5356	Mg, Mn, Cr	Weld wire and electrodes
6201	Si, Mg	Electrical
6262	Mg, Si, Cu, Cr, Pb, Bi	Screw machine products
8176	Fe, Si	Electrical

15.2. AUSTENITIC STAINLESS STEELS

15.2.1 Nature and general importance

The term "stainless steel" generally refers to an alloy with at least 10.5% chromium content that exhibits minimal corrosion in normal atmospheric exposure. Such alloys may not be immune to corrosion in more aggressive environments, on the other hand, properly alloyed stainless steels can resist many of these aggressive environments. The resistance of stainless steels to corrosion arises from the tightly adhering, relatively impervious chromium-oxygen layer that forms on the alloy surface.

Most commercial stainless steel is the *austenitic type* that has FCC austenite structure at room temperature. This is in contrast to carbon steels, where austenite may not form unless the steel is heated above 732°C, and where 1000°C may be recommended for full austenitization, as discussed in Chapter 14. With most austenitic stainless steels, the austenite is made "stable" at much lower temperatures by the addition of nickel in amounts from 6 to 22%, and most of these steels retain their austenitic structure throughout the cryogenic range.

The austenitic stainless steels are quite tough, with many alloys displaying no ductile-to-brittle transition. They are ductile, and generally display a high work-hardening rate, allowing very high strengths to be achieved through cold work. On the other hand, such toughness and work hardening may be a hindrance for many machining and cold forming operations.

Primary processing involves general steelmaking technologies, although great care must be exercised to avoid loss of chromium due to oxidation.

15.2.2 Flow stress, cold work, and annealing

A stress-strain curve for 304 stainless steel (18% Cr, 8% Ni) is illustrated in Figure 11.8. Using the relationship $\sigma_o = k\varepsilon_o^N$, the stainless steel curve in Figure 11.8 can be approximated by

$$\sigma_o(MPa) = 1130\,\varepsilon_o^{0.36}. \tag{15.4}$$

As a word of precaution, equations such as Equation 15.4 should be based on actual data for the material sample of interest. Austenitic stainless steel flow curves vary greatly, particularly when there is partial transformation of austenite to martensite during the deformation. Austenitic stainless steels displaying this behavior are called "unstable." Lower nickel content promotes such austenite instability.

In addition to the obvious dependence on strain, austenitic stainless steel flow stress is dependent on temperature and strain rate. The effect of strain rate for 304 stainless steel, $d\varepsilon_o/dt$, based on various data in the literature, is reasonably expressed by Equation 15.5:

$$\sigma_o = G(d\varepsilon_o/dt)^{0.014}, \qquad (15.5)$$

where G is the strain rate strength coefficient. The effect of temperature on this alloy is reasonably expressed by Equation 15.6:

$$\sigma_o(MPa) = \sigma_{To} + (-2.4)(T-T_o), \qquad (15.6)$$

where σ_{To} is the flow stress in MPa at the reference temperature T_o, in °C. Figure 15.2 displays some calculated relationships among flow stress, strain rate, and temperature for type 304 stainless steel.

Austenitic stainless steel alloys comparable to type 304 are typically annealed in the 1000 to 1120°C range, followed by rapid cooling to avoid chromium carbide precipitation.

15.2.3 Grades of austenitic stainless steels

The most common stainless steel, by far, is grade 304 (UNS Designation S30400), with a chromium range of 18–20% and a nickel range of 8–10.5%.

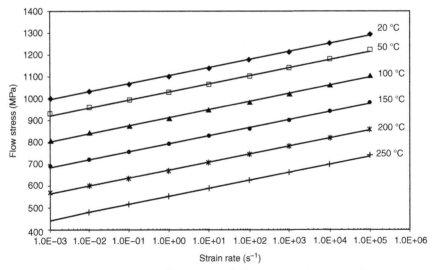

Figure 15.2 Calculated relationships among flow stress, strain rate, and temperature for 304 stainless steel. From C. Baid and R. N. Wright, *Wire Journal International*, 37(3), 2004, 66.

Commercial alloy content will often be near the lower end of such ranges, and the term "18–8" is used to describe 304 and similar alloys. The tensile strength of grade 304 is at least 515 MPa in annealed form, but it is readily cold worked to twice this level. Carbon levels are generally low in austenitic stainless steels, with 0.08% commonly cited for grade 304. Lower carbon levels may be beneficial to as-welded corrosion resistance, and the grade 304L (UNS Designation S30403), with 0.03% C, is available as a low-carbon alternative to grade 304.

The corrosion resistance of the austenitic stainless steels, particularly pitting resistance, can be considerably enhanced with the addition of 2–3% molybdenum. Grades 316 (UNS Designation S31600) and 316L (UNS Designation S31603) are molybdenum-bearing alternatives to grades 304 and 304L (involving somewhat lower chromium and somewhat higher nickel contents).

Austenitic stainless steels are notoriously difficult to machine, and grades 303 (UNS Designation S30300) and 303Se (UNS Designation S30323) have been developed for applications where machinability is a particularly serious consideration. Grade 303 involves an addition of at least 0.15% sulfur, and grade 303Se involves an addition of at least 0.15% selenium, in addition to 0.06% sulfur. The soft sulfides and selenides present in the microstructure help break up chips and generally facilitate cutting and ductile fracture. On the other hand, the presence of these phases in the microstructure often lowers corrosion resistance, fatigue strength, and ductility.

The high work-hardening rates of the austenitic stainless steels may become even higher in the case of certain lower nickel and chromium variations. This is because some of the austenite will transform to martensite during deformation. Grade 301 (UNS Designation S30100) is a common alloy displaying this characteristic, available at minimum tensile strength levels of 1275 MPa while still manifesting 9% elongation. High rates of work hardening can complicate some cold forming operations. In such cases, grade 305 (UNS Designation S30500) has been used as an alternative to grade 304, providing about 10% lower strength with increased nickel content.

Mid-twentieth century concerns over nickel supply led to the substitution of manganese and nitrogen for nickel, resulting in the 2xx grades. Such grades are still in use. Grade 201 (UNS Designation S20100) is an example, containing 16–18% chromium, 3.5–5.5% nickel, 5.5–7.5% manganese, and 0.25% nitrogen with mechanical properties similar to grade 304.

15.3. TOOL STEELS

15.3.1 Nature and general importance

While the term "tool steel" can be rather broad and generic, the basic idea is that these steels provide high or very high strength and wear resistance, with a reasonable trade-off with toughness and shock resistance. For applications involving rapid, repeated loading, high fatigue strength is required. Where appropriate, these properties may extend to high temperatures, including resistance to thermal shock and thermal fatigue. Tool steel rod and wire are precursors to tool bits and saw teeth as well as other products.

15.3.2 Processing and phase relations

The general metallurgy of these steels involves substantial carbon content, and high, or very high, content of stable carbide forming elements, such as chromium, molybdenum, tungsten, and vanadium. The end product is a tempered martensite containing stable carbides that resist over-tempering or softening during use.

Tool steels are prepared as discrete ingots, prior to hot deformation en route to bar and rod form. The properties of these steels improve with microstructural refinement and homogeneity, and fine-structured, powder metallurgically derived tool steels have been on the market in modern times. Figure 15.3 illustrates steps in the preparation of a tool steel ingot by powder metallurgy processing.

Tool steels are best drawn in a spheroidized condition, with a limited number of reductions possible prior to process annealing. Workability is limited. Hardening and tempering are final stages in the overall product processing.

15.3.3 Flow stress, cold work, and annealing

Due to the limited drawability of tool steels, without process annealing, development of an extended flow curve is impractical. Moreover, the flow stresses of stock before and after a given pass may be similar. Figure 15.4 shows an engineering stress-engineering strain curve for an annealed M42 tool steel. Figure 15.5 displays the limited fracture strain and the "quasi-cleavage" mode of fracture. Drawing pass reductions can, however, involve strains well beyond those achievable in a tensile test.

Process annealing temperatures for tool steel grades are often in the 850–900°C range, although specific recommendations should be obtained from the vendor. The cooling rate from the annealing temperature is

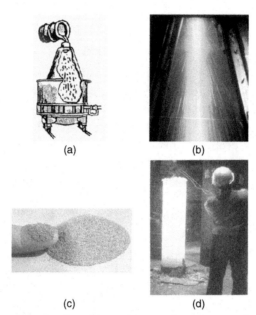

(a)   (b)

(c)   (d)

Figure 15.3 Processing of tool steel ingots by way of powder metallurgy: (a) schematic of operation, (b) the atomization process, (c) resultant powder, and (d) sintered ingot. (Courtesy of Crucible Industries, LLC).

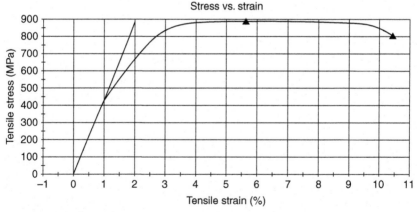

Figure 15.4 Engineering stress strain curve for an annealed 42 tool steel.

important and rates of 10–22°C per hour are often recommended, at least at higher temperatures.

15.3.4 Grades of tool steels

Although vendors often market tool steels under proprietary trade names, nearly all of these products are described within the AISI or UNS grade

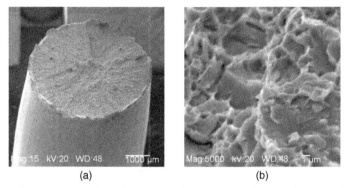

(a)   (b)

Figure 15.5 Tensile test fracture surface detail for an M42 tool steel displaying: (a) limited fracture strain and (b) a quasi-cleavage mode of fracture.

systems. In the AISI system, the grades often begin with a letter that roughly describes the application or composition of the steel: M contains large molybdenum additions, T contains large tungsten additions, H for hot work, A for air hardening; O for oil hardening, W for water hardening, and S for shock resisting. The term "high speed steel" refers to material suited for high speed cutting tool applications. Grades beginning with M and T are in this category. There are additional categories, including the important D grades, which contain large chromium additions and are designed for cold working. Powder metallurgically processed grades often begin with PM, and UNS grade system tool steels begin with T.

The tool steel grade spectrum involves a trade-off between toughness, or impact/shock resistance, and wear resistance or hardness. For good toughness, with somewhat modest wear resistance, S-7 (T41907 in the UNS system) is used. This grade is a robust Cr-Mo-V alloy steel with about 0.5% carbon. For increased wear resistance, the grades A2 (T30102) and D2 (T30402) have been used for many years. Grade A2 is also a robust Cr-Mo-V alloy, but with about 1.00 % carbon. Grade D2 can be called a Cr-Mo-V alloy, but the alloy content is high, including about 12% chromium with 1.5% carbon.

Grades M1 (T11301) and M2 (T11302) are widely used for cold forming. The carbon contents of these steels approach 1.0%. In the case of M1, a molybdenum addition of about 8.5% is combined with major additions of chromium, vanadium, and tungsten. In the case of M2, the molybdenum content is lowered in conjunction with increased tungsten and vanadium additions. For high performance, high wear resistance applications, the powder metallurgical PM-M4 is available. The M4 (T11304) composition contains increased carbon and vanadium, in comparison to M2. The powder metallurgical grade PM-A11 provides exceptionally high wear resistance, typically containing nearly 10% vanadium with 2.45% carbon, 5.25% chromium, and 1.35% molybdenum.

It is important to understand that the performance of these steels can be manipulated with tempering practice, and that higher hardness will usually correlate with increased wear resistance and diminished toughness. The high wear resistance alloys cited above are often tempered to Rockwell C scale hardnesses in the 62 to 65 range, whereas Rockwell C scale hardnesses well down into the 50s correspond with relatively high toughness.

Regarding hot working tool steels, there is wide use of the H10 (T20810), H12 (T20812), and H13 (T20813) alloys. These grades are of the Cr-Mo-V type with medium carbon contents in the 0.30–0.45% range. Grade H12 also includes tungsten.

15.4. NICKEL AND NICKEL ALLOYS

15.4.1 Nature and general importance

Nickel is an FCC metal with corresponding toughness, ductility, and work-ability. Nickel and nickel-based alloys have excellent corrosion and oxidation resistance combined with high-temperature strength, which is needed in high temperature wire applications such as with nickel-chromium alloys for resistance heating.

Sulfide deposits are a major source of nickel for primary processing, particularly with nickel-copper-iron sulfide ores called pyrrhotite. With appropriate comminution, magnetic separation, and froth flotation technology, the nickel ore may be isolated from that of iron and copper, with copper and iron as the important by-products of the nickel refining process. Concentrated nickel-bearing ore is roasted, smelted, and converted to a *matte* containing nickel and copper sulfides. Upon cooling, the matte forms nickel and copper sulfides and nickel-copper alloy. The nickel sulfide is roasted to form nickel oxides, which can be further processed to nickel and nickel alloys.

15.4.2 Flow stress, cold work, and annealing

Figure 15.6 displays a flow curve for nickel wire.[114] Using the relationship $\sigma_o = k\varepsilon_o{}^N$, the nickel curve in Figure 15.6 can be approximated by

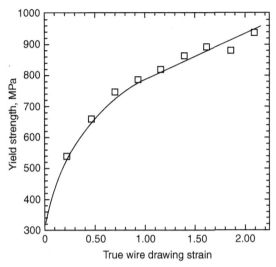

Figure 15.6 Flow stress curve for nickel wire. From S. K. Varma, *Metallurgy, Processing and Applications of Metal Wires*, H. G. Paris and D. K. Kim, editors, The Minerals, Metals and Materials Society, Warrendale, PA, USA, 1996, 19. Copyright held by The Minerals, Metals and Materials Society, Warrendale, PA, USA.

$$\sigma_o(\text{MPa}) = 790 \, \varepsilon_o^{0.25}. \tag{15.7}$$

Again, as a word of precaution, equations such as Equation 15.7 should be based on actual data for the material sample of interest.

In addition to the obvious dependence on strain, nickel flow stress is dependent on temperature and strain rate as well. The effect of strain rate for nickel, $d\varepsilon_o/dt$, based on the ambient temperature work of Follansbee et al.,[115] is expressed by Equation 15.8:

$$\sigma_o = G(d\varepsilon_o/dt)^{0.012}, \tag{15.8}$$

where G is the strain rate strength coefficient. The effect of temperature on nickel in the drawing temperature range, based on the work of Pasquine et al.,[116] is expressed by Equation 15.9:

$$\sigma_o(\text{MPa}) = \sigma_{To} + (-0.008)(T - T_o), \tag{15.9}$$

where σ_{To} is the flow stress in MPa at the reference temperature T_o, in °C. Data for several other nickel-base compositions are summarized in Table 15.5. The effects of temperature on flow stress are relatively low, and in the case of nominally pure nickel there is little decrease of flow stress with temperature in the range of interest for wire drawing analysis.

Table 15.5 Flow curve representation information (MPa) for five nickel-base alloys[117]

Composition	Strain Rate Effect	Temperature Effect
Ni510: 0.05% C	$\sigma_o = G (d\varepsilon_o/dt)^{0.0069}$	
Ni200: 0.08%C		$\sigma_o = \sigma_{To} + (-0.054) (T - T_o)$
Ni1900: 0.19% C	$\sigma_o = G (d\varepsilon_o/dt)^{0.0072}$	
Duranickel 301: 4.4% Al, 0.6% Ti		$\sigma_o = \sigma_{To} + (-0.28) (T - T_o)$
80Ni-20Cr	$\sigma_o = G (d\varepsilon_o/dt)^{0.0145}$	

From R. N. Wright, C. Baid and K. Baid, Wire Journal International, 39(3) (2006) 146.

Table 15.6 Nickel and nickel alloy grades used in wire form

Designation Number	Principal Alloying Elements	Applications
Nickel 200, 201, 205	None	Electronic parts, lead-in wires, support wires
Nickel 211	4.8% Mn, 0.4%Fe	Sparking electrodes, support wires, winding wires
Duranickel 301	4.4% Al, 0.6% Ti	Springs, clips
Alloy W	24% Mo, 6% Fe, 5% Cr, 2.5% Co	Welding of dissimilar alloys
80 Ni–20 Cr	20% Cr	Resistance heating
65 Ni–15 Cr– 20 Fe	20% Fe, 15% Cr	Resistance heating

15.4.3 Nickel and nickel-alloy grades important to wire technology

A number of nickel and nickel alloy grades used in wire form are summarized in Table 15.6.

15.5. QUESTIONS AND PROBLEMS

15.5.1 An aluminum member is labeled 7075-T6. What can be learned about this product from Tables 15.2 and 15.3?

Answer: Its principal alloying element is zinc, it has been solution treated, and it is allowed to artificially age.

15.5.2 An austenitic 304 stainless steel rod is tensile tested in the laboratory at 20°C with a strain rate of 10^{-3} s^{-1}. It is to be drawn at a strain rate of 10^2 s^{-1} with thermomechanical heating to about 150°C. How accurate is the laboratory strength data for drawing analysis?

Answer: Consulting Figure 15.2, the flow stress under the laboratory testing conditions should be about 1000 MPa. Under drawing conditions, the flow stress should be about 850 MPa. Use of the laboratory data will clearly overestimate the flow stress under drawing conditions.

CHAPTER SIXTEEN

Wire Coatings

Contents

16.1. REASONS FOR COATING WIRE

In many applications wire is often, or nearly always, used with a coating. This is done because of the need for electrical insulation and corrosion protection. Other important motivations include the need to provide a superior surface for lubrication and to facilitate bonding to other structural media, such as the role of a brass coating in bonding steel to rubber.

This chapter reviews the rudiments of important coating technologies, where such coatings are a prominent feature of wire in a semi-finished product form. This chapter will not, however, review coatings that play transient roles in wire processing, such as lubricants, lubricant carriers, cleaners, and temporary corrosion inhibitors. Aspects of these latter coatings are discussed in Chapters 8 and 17.

16.2. COATING TYPES AND COATING PROCESSES

Coatings for electrical insulation include *enamels* and *polymers* of many types. Coatings for corrosion protection include *metals* such as zinc and

Wire Technology
ISBN 978-0-12-382092-1, DOI: 10.1016/B978-0-12-382092-1.00016-6

aluminum, as well as polymers. Lubrication and bonding may be facilitated with copper and copper alloys. Paper wrapping also serves a useful purpose.

Major coating technologies include enameling, extrusion, hot dipping, electro–coating, electro–less coating, cladding, and wrapping.

16.3. ENAMELS AND ENAMELING
16.3.1 General logistics

Enamels are most prominently associated with copper and aluminum magnet wire. These smooth, thin insulators minimize space consumed in wound coils, have high dielectric strength, and may provide moisture and solvent resistance.

A coating of enamel, in a resinous solution, is applied to the wire, and the wire passes through a die so as to leave a film coating on the wire. This initial coating is dried and cured, and the process is repeated as many as six times. Figure 8.21 reveals these coatings in cross section. The coatings may involve more than one enamel formulation. Overall increase in wire diameter attributable to the layers of enamel is called the *build*. Build specifications may be described as "single," "heavy," and "triple," where a heavy build is twice the thickness of a single build, and a triple build is three times the thickness of a single build.

Facilities for enameling are called enameling "towers," because of the vertical layout of drying and curing operations. Enameling towers may be in tandem with drawing, annealing, and cleaning operations.

16.3.2 Types of enamels

A practical listing and general discussion of major magnet wire enamels or dielectrics is presented in the *Electrical Wire Handbook*.[118] Enamels may be used in combination, and numerous variations exist. A brief listing follows:

Plain enamel. Synthetic resins and modifying agents. Applications include transformer and telephone coils.

Nylon. A tough thermoplastic that may be removed by molten solder. Applications include automotive electrical devices.

Polyurethane. Formed by reacting a form of isocyanate and a polyester. May be removed with molten solder. Used in fine wire applications.

Polyvinyl formal (Formvar). Classical enamel (see Figure 8.21), involving a reaction with phenolformaldehyde and possibly with isocyanate and melamine. Requires stress relief. Is still used where resistance to oil may be required.

Solderable acrylic. Water dispersion of acrylonitrile polymer, acrylic acid, and butyl acrylate is applied. May be removed by molten solder. Applications include automotive electrical devices.

Epoxy. Early resin solution coatings had properties similar to those of polyvinyl formal. Powder coating technologies have subsequently become available. Provides good resistance to oil containing media.

Polyester. This enamel is from the terephthalic acid category. Good heat and heat shock resistance. Used in a variety of electrical and electrical coil products.

Polyimide. Involves the reaction of pyromellitic dianhydride with an aromatic diamine. Very high thermal stability and overload resistance. Resistant to solvents and acids, but not to alkalis. Used in high performance electrical devices.

Polyester-polyamide-imide. The base is a modified polyester with an overcoat of amide-imide polymer. Superior properties and wide general application.

16.4. EXTRUSION

16.4.1 General logistics

In the polymer extrusion process (not to be confused with metal deformation processing, such as aluminum extrusion), a thermoplastic polymer in particulate form is fed into a hopper, and it flows into a screw-containing barrel. The action of the screw drives the powder ahead, achieving melting and pressurization of the polymer. Figure 16.1 provides a schematic illustration of a typical screw extruder.[119] For wire coating applications, the pressurized, molten polymer is forced into a crosshead where it surrounds heated wire stock, which then becomes coated as its passes through the crosshead and out a die. A schematic illustration of the crosshead is displayed in Figure 16.2.[120]

16.4.2 Polymer coatings

A practical listing and general discussion of major polymers for wire coating can be found in the *Ferrous Wire Handbook*.[121] A brief listing follows.

Softened polyvinyl chloride (PVC). Since PVC can be hard and brittle at ambient temperature, softeners, such as esters, are added for wire

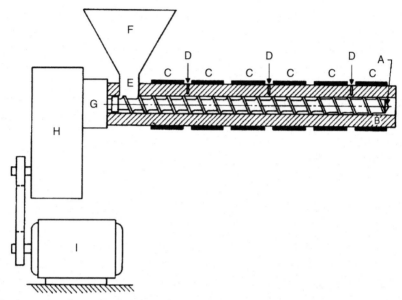

Figure 16.1 Schematic drawing of a polymer screw extruder: A, screw; B, barrel; C, heater; D, thermocouple; E, feed throat; F, hopper; G, thrust bearing; H, gear reducer; I, motor. From P. N. Richardson, *Introduction to Extrusion*, Society of Plastics Engineers, Brookfield Center, CT, USA, 1974, 4. Copyright held by Society of Plastics Engineers, Brookfield Center, CT, USA.

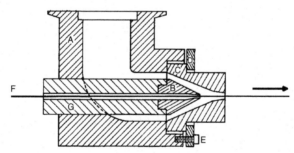

Figure 16.2 Schematic drawing of a wire-coating pressure die: A, die body, cross head; B, guider tip; C, die; D, die retaining ring; E, die retaining bolt; F, wire; G, core tube. From P. N. Richardson, *Introduction to Extrusion*, Society of Plastics Engineers, Brookfield Center, CT, USA, 1974, 84. Copyright held by Society of Plastics Engineers, Brookfield Center, CT, USA.

coating applications. Stabilizing compounds and filler material may be added to the softened PVC. Polyvinyl chloride is a good electrical insulator.

Polyamide plastics, such as nylon. Provides high strength and toughness.

Polyethylene. Highly resistant to solvents, water, and many caustic salt solutions and organic acids.

Polypropylene. Strength, toughness, and chemical resistance are generally greater than for polyethylene, but polypropylene is vulnerable to daylight exposure.

Polycarbonate. Strength, toughness, and chemical resistance are generally high, but polycarbonate is vulnerable to alkaline media, amines, and ammonia solutions.

Extrudable polyfluorocarbons. Some major polyfluorocarbons, including Teflon® are not extrudable; however, family members such as polyfluorochloroethylene are. These compounds have strong resistance to essentially all acids and alkalis. Extrudable polyfluorocarbons are less stable at elevated temperature than Teflon®.

16.5. HOT DIPPING

With this technology, a coating is achieved by passing the wire through a bath of molten coating material. Upon emerging from the bath, the wire is passed through a wiping member that trims the coating to a desired thickness. The role of the wiping process is somewhat analogous to that of the enameling die cited earlier. The dip coating procedure may be preceded by a thermal treatment, by pickling and cleaning, and by application of a flux. It may be followed by annealing, quenching, and supplemental coating. In some cases, a metallurgical reaction between the coating and the substrate is brought about by the hot dipping action or a subsequent anneal. A prominent example is galvannealing, where intermetallic compounds of iron and zinc are formed between the zinc and the iron. Supplemental coatings may include waxes, polymers, and dichromates (where permissible). A schematic illustration of a hot-dip coating line is presented in Figure 16.3.[122]

Hot dipping technology has been widely applied to the coating of steel with zinc (hot-dip galvanizing), and practices have been developed for zinc alloys, such as Zn-Al and Zn-Al-Sn. Aluminum may be applied by hot dipping. Zinc and aluminum coatings are generally for corrosion resistance, and the coating acts as a sacrificial anode relative to the steel substrate. Tin may be applied to steel or copper with hot-dip technology.

16.6. ELECTRO-COATING

In this technology, the wire, as a cathode, is exposed to an electrolyte containing cations of a coating metal. The cations of the coating metal are

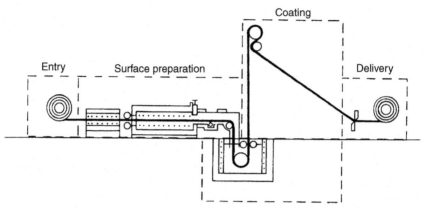

Figure 16.3 Illustration of a classic hot-dip coating line. From A. R. Cook, Galvanized Steel, *Encyclopedia of Materials Science and Engineering, Vol. 3*, M. B. Bever, editor-in-chief, Pergamon Press, Oxford, UK, 1986, 1899. Copyright held by Elsevier Limited, Oxford, UK.

reduced to the coating metal at the cathode. The coating ions may enter the electrolyte at the anode. Electroplating is a versatile technology, applicable to many systems. Common plating materials include zinc, chromium, nickel, cadmium, copper, tin, aluminum, and various noble metals (gold, silver, platinum, etc.).

In the classic example of electrogalvanizing, a zinc anode is the source of zinc ions that migrate to the steel wire cathode, where they are reduced to metallic zinc. In principle, the process can be used on any steel, and little reaction occurs between the zinc and the iron.

Tire cord wire is electroplated with brass to facilitate adhesion of the steel wire to rubber. In one such technology, referred to as *diffusionless*, brass is plated from an electrolyte (typically sodium cyanide solution) containing both copper and zinc cations. An alternate technology is *thermodiffusion*, where copper is electroplated followed by zinc. A subsequent thermal treatment allows the copper and zinc to interdiffuse en route to the brass alloy.

Tire bead wire may be electroplated with bronze using diffusionless technology. A stannate–cyanide bath is used with a few percent tin plating together with copper.

Further applications include copper on aluminum in electrical assemblies, tin plating of copper to ease soldering, precious metal plating of electronic circuit materials (to avoid oxidation and increased contact resistance), and various coatings for aesthetic purposes.

16.7. OTHER COATING TECHNOLOGIES

Apart from enameling and extrusion, polymer coatings may be applied by *fluidized bed* technology and *immersion*. In the fluidized bed approach, heated wire is passed through powdered plastic that is "fluidized" with gas flow. The plastic powder melts on contact with the heated wire, creating a coating of appropriate thickness with proper process design and control. Immersion technology may be subdivided into *hot* and *cold* processes. In the hot approach, a heated wire is immersed in a plastic paste, which forms the desired coating. In the cold approach, the wire is not heated before immersion in a plastic paste. However, the plastic must be cured in a subsequent thermal treatment.

In some applications magnet wire is wrapped with paper or tape insulation. Glass fibers and polyester-glass "yarns" also may be applied by wrapping.

Metal coating or cladding may be physically applied to a wire substrate and then drawn as a composite. For example, a rod may be placed in a tube, and then the two-piece assembly may be drawn or extruded. Analogous techniques involve casting of the coating metal on a rod or billet substrate and hot-isostatically pressing coating metal(e.g., powder) on a rod or billet substrate.

Coating applications involving spraying and vapor deposition also have been developed.

16.8. ZINC ALLOY COATING OF STEEL – A DETAILED ILLUSTRATION AND ANALYSIS[123]

A zinc-5% aluminum alloy may be usefully applied to a steel substrate by way of the Galfan® process. A double-dip Galfan® line is schematically illustrated in Figure 16.4. In the double-dip method, the first dip involves a conventional hot-dip galvanization process, followed by a second hot-dip in the Zn-5%Al alloy. This double-dip approach is necessary because of the incompatibility of Zn-5%Al with many flux systems.

The galvanized coating acquired in the first dip generally forms a sequence of intermetallic iron-zinc compounds between the iron substrate and an outside layer of nearly pure zinc. The intermetallic compounds are embrittling. However, upon passage through the hot Galfan® bath, the coating from the prior bath is transformed into an aluminum-iron-zinc intermetallic compound. The final coating involves a layer of Galfan® alloy on top of the layer of Al-Fe-Zn intermetallic compound.

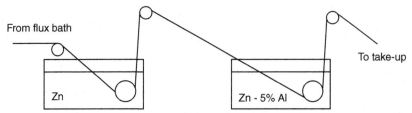

Figure 16.4 Schematic illustration of a double-dip Galfan® line. From F. E. Goodwin and R. N. Wright, *The Process Metallurgy of Zinc-Coated Steel Wire and Galfan® Bath Management, Conference Proceedings*, Wire & Cable Technical Symposium, 71st Annual Convention, The Wire Association International, Inc., Guilford, CT, 2001, 135.

The aluminum that enters the Al–Fe–Zn intermetallic depletes the aluminum composition of the Galfan® alloy bath, thus requiring careful monitoring and management of this bath. The following model of Galfan® bath aluminum concentration is useful for bath management.

If the process line is able to add Galfan® alloy to keep the bath mass constant, then the incremental depletion of aluminum associated with the increment of wire coating can be expressed as:

$$M\,(dC_{Al}) = -M_{Al}\,(dA) + (0.05)\,M_G\,(dA) - C_{Al}\,M_{Go}\,(dA), \quad (16.1)$$

where C_{Al} is the aluminum concentration in the bath, dC_{Al} is the incremental change in aluminum bath concentration, dA is the increment of wire surface coated, M is the mass of the bath, M_{Al} is the mass of aluminum per unit of wire surface area in the transformed intermetallic layer, M_G is the net mass of Galfan® alloy melt removed per unit of wire surface area, and M_{Go} is the mass of the Galfan® alloy overlay that solidifies per unit of wire surface area.

The left-hand side of Equation 16.1 is the incremental mass of aluminum lost when an increment of wire surface area is coated in the Galfan® alloy bath. There are three contributions to the incremental mass, specifically the three terms on the right side of Equation 16.1. The first term is the mass of aluminum lost to the intermetallic layer. The second term is the amount of aluminum added to the bath from the replenishment process. The third term is the amount of aluminum lost due to solidification of the bath alloy on the wire.

Equation 16.1 can be reorganized and integrated to yield Equation 16.2.

$$C_1 = (Z + C_0)\,\exp(-YA) - Z \quad (16.2)$$

where C_0 is the concentration of aluminum in the starting bath, C_1 is the concentration of aluminum in the bath after coating a surface area of A, Y is (M_{G0}/M), and Z is $[1 - (0.05)\ M_G]/M_{G0}$.

Using Equation 16.2 and the practical values of descriptors and parameters involved, bath maintenance practices can be planned. The general approach would be to associate a rate of Galfan® alloy bath replenishment with a rate of supplemental aluminum addition. Periodic bath analysis should be undertaken. Monitoring of bath replacement needs can provide improved estimates of M_G, the galvanized coating weight, and the overall coating weight. The value of M_{G0} can be estimated as 20 g/mm² less than the overall coating rate. With process experience and observations, increasingly sophisticated bath management practices will evolve.

16.9. COMPOSITE MECHANICAL PROPERTIES OF COATED WIRE

In many cases strength and ductility specifications for coated wire are based on measurements of the bare wire. While this may suffice in an empirical way for certain commercial quality purposes, bare wire mechanical properties vary considerably from those of the composite, coated wire. This is especially the case where bending and twisting are involved, since these deformation modes develop maximum strain at the composite wire surface where coating properties may dominate. The contribution of the coating to composite mechanical properties is also substantial in tensile deformation, as illustrated by the work of Wright and Patenaude.[124]

Wright and Patenaude conducted tests on PVC insulated, 8 AWG copper wire, revealing the respective contributions of the PVC coating and the core wire to strength in tension and to elongation. The presence of the PVC jacket was observed to forestall necking and increase tensile elongation when the composite was compared to the copper. The tensile properties measured for the copper are illustrated in Figure 16.5, and those for the PVC alone are seen in Figure 16.6. Tensile test results for the composite, coated conductor are shown in Figure 16.7. In these tests, stress-strain relations after necking or at high strains must be interpolated between the initial portion of the stress-strain curve and values measurable at necking or fracture. In any case, it is clear that the composite conductor has higher uniform elongation and fracture strain than the copper on a stand-alone basis.

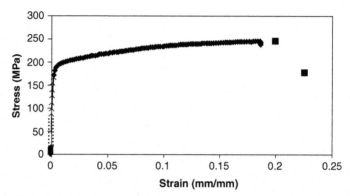

Figure 16.5 Engineering stress-strain curve for copper removed from its PVC insulation. From R. N. Wright and K. Patenaude, *Wire Journal International*, 34(5) (2001) 94.

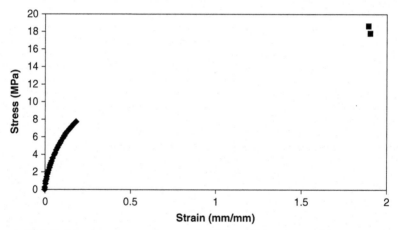

Figure 16.6 Engineering stress-strain curve for PVC insulation removed from copper wire. From R. N. Wright and K. Patenaude, *Wire Journal International*, 34(5) (2001) 94.

The strength of the composite, in tension, is less than that of the copper on a stand-alone basis. The simplest projection of composite properties in tension is the "rule of mixtures," as per:

$$\sigma_{\alpha+\beta} = \sigma_\alpha \left(A_\alpha/A \right) + \sigma_\beta \left(A_\beta/A \right), \qquad (16.3)$$

where $\sigma_{\alpha+\beta}$ is the strength of the composite, or α and β, in tension; σ_α is the strength of component α; σ_β is the strength of component β;

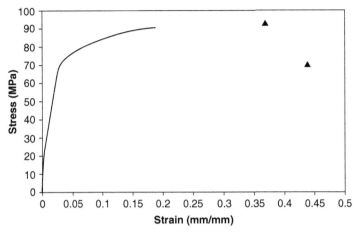

Figure 16.7 Engineering stress-strain curve for PVC insulated 8 AWG copper conductor. From R. N. Wright and K. Patenaude, *Wire Journal International*, 34(5) (2001) 94.

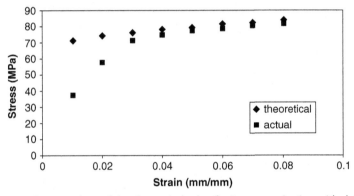

Figure 16.8 A comparison of the theoretical rule of mixtures projection with data from Figure 16.7. From R. N. Wright and K. Patenaude, *Wire Journal International*, 34(5) (2001) 94.

(A_α/A) is the cross-sectional area fraction of α; and (A_β/A) is the cross-sectional area fraction of β. A rule of mixtures projection from the data of Figures 16.5 and 16.6 is compared with the composite data of Figure 16.7 in Figure 16.8. Equation 16.3 is consistent with the actual composite tensile data at strains beyond 0.03 to 0.04 (3–4%). Below this strain range, the rule of mixtures overestimates the strength of the composite. It is hypothesized that this discrepancy is related to the low interfacial strength between the copper and the PVC.

16.10. QUESTIONS AND PROBLEMS

16.9.1 Equation 16.1 involves three components of the incremental mass of aluminum lost (or gained) when an increment of surface area is coated in the Galfan® process. Reviewing Figure 16.4, make certain that you understand the aluminum transfer involved with the three components. Using a sketch of Figure 16.4 and arrows, make a diagram of the transfers of aluminum.

Answer: The first term involves a transfer of aluminum into the iron–zinc coating that was established in the zinc bath. This first term aluminum transfer actually occurs in the second, Zn-5%Al bath. The second term involves aluminum directly added to the Zn-5%Al bath. The third term involves aluminum lost as the Zn-5%Al bath alloy solidifies on the wire.

16.9.2 Consider Figures 16.5 and 16.7. What are the implications of the coating contribution to tensile properties for applications that involve winding under tension?

Answer: The coated, composite conductor yields at a significantly lower stress than copper by itself (70 vs. 180 MPa). On the other hand, the coated, composite conductor may be elongated considerably more than the copper alone (strains of 0.35 vs. 0.18).

CHAPTER SEVENTEEN

Redraw Rod Production

Contents

17.1. THE ROD ROLLING PROCESS

While limited quantities of redraw stock are prepared by extrusion, swaging, cogging, and bar drawing, the vast majority of redraw rod emerges from a rod rolling operation, which will be detailed in this chapter.

17.1.1 The issue of spreading in rod rolling

The ultimate objective of redraw rod rolling is progressively lengthening the rod and reducing the cross-sectional area to a dimension that permits drawing. This is not a simple objective, as rolling from a round to a round in a single pass cannot happen because of side spread. That is, upon rolling, a circular cross section will decrease in dimension in the direction perpendicular to the roll axis, and will increase in dimension (or spread) in the direction parallel to the roll axis. Thus it is necessary to roll a sequence of alternate cross section shapes, such as round-to-oval-to-round, and so forth. In this pursuit, dimensional control issues are primarily related to difficulties in predicting and controlling side spread in the individual passes. Too little spread produces incorrect geometry for the next pass, and too much spread creates "fins," "flashing," "cold shuts," and so on.

Wire Technology
ISBN 978-0-12-382092-1, DOI: 10.1016/B978-0-12-382092-1.00017-8

17.1.2 A simple approach to spread estimation

Arnold and Whitton have discussed a useful approach to spread prediction in the absence of detailed historical data.[125] The key to this approach is the expression for the ratio of final to initial widths, (W_1/W_0)

$$W_1/W_0 = n_s(h_0/h_1)^{1/2}, \qquad (17.1)$$

where h_0 is the height (perpendicular to the roll axis) of a rectangle of width W_0, with an area equal to the rod cross-sectional area before the pass; where h_1 is the height of a rectangle of width W_1, with an area equal to the rod cross-sectional area after the pass; and where n_s is an empirical constant dependent on rod flow and friction, among other factors. The relation of W and h to basic round and oval geometry is illustrated in Figure 17.1.

The value of h_0 is A_0/W_0 and the value of h_1 is A_1/W_1, where A_0 and A_1 are, respectively, the rod cross-sectional areas before and after the pass. Substitution of these values into Equation 17.1 leads to

$$W_1/W_0 = n_s^2 (A_0/A_1), \qquad (17.2)$$

which indicates that the spreading can be regarded as proportional to the cross-sectional area ratio. The relationship between A_0 and A_1 may be a pass design parameter; for example, a given area reduction or apparent true strain per pass may be specified. In many instances, Equation 17.2 is more convenient than Equation 17.1, although Equation 17.1 represents a better picture of the relationship between thickness reduction and spreading, and more clearly indicates the basis for the relationships among the various geometries.

In practice the value of n_s ranges from just below 0.9 to over 1.0; higher values indicate a greater tendency for spread. The theoretical lower limit of n_s is dictated by a condition of no spreading, or $(W_1/W_0) = 1$. Substituting

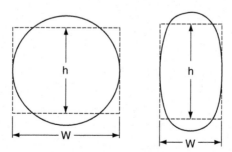

Figure 17.1 Basic round and oval geometries and the relationship of W and h.

this into Equation 17.2 leads to a minimum value for n_s of $(1-r)^{\frac{1}{2}}$, where r is the reduction in the pass in decimal form. The theoretical maximum value of n_s would correspond to a condition of no area reduction, or $(A_0/A_1) = 1$. In this case, Equation 17.2 indicates that the value of n_s would be $(W_1/W_0)^{\frac{1}{2}}$, or limited only by the magnitude of the reduction. Large values of n_s are encountered with large reductions. They are also encountered in the rolling of tubes and certain composite rods with powder cores because of the ease of cross-sectional buckling.

Practical values of n_s must be deduced from preliminary rolling trials or from past, successful roll groove designs. Moreover, n_s will vary with pass geometry, line tension, rolling temperature, friction conditions, and rod process history. Nevertheless, n_s is a relatively simple guide to expected rod deformation under conditions of a given pass, and can be the basis for the design of roll groove geometry, for determination of roll gaps, for rolling process troubleshooting, and for rolling process correction. The author has published a comprehensive discussion of the practical analysis of roll pass design.[126]

17.2. THE OVAL/ROUND SEQUENCE

17.2.1 Basic relationships

Rolling pass schedules leading to round stock generally employ sequences of ovals and rounds, such as illustrated in Figure 17.2.[127] The alternating geometry ensures that the width of the entering rod section is less than the width of the emerging rod section, which is necessary for roll entry. As a practical matter, actual rod cross sections may be "out of round" and not rigorous ovals. Nevertheless literal oval and round geometry is employed in analysis. In general, it is useful to consider oval/round passes in pairs, starting with a round, going to an oval, and going back to a round.

It is useful to pursue this objective literally. Considering the first pass in terms of Equation 17.1, the value of w_0 is D, the diameter of the entering round. Hence, A_0 is $(\pi/4)D^2$, and h_0 is A_0/W_0, or $(\pi/4)D$. This first pass produces an oval with a major diameter B and minor diameter, b, where W_1 is B. Hence, A_1 is $(\pi/4)Bb$ and h_1 is A_1/W_1 or $(\pi/4)b$. The ratio A_1/A_0 will be $(1-r)$.

Now let us consider the second pass, going from the oval back to a round, in terms of Equation 17.1. The value of W_0 will be b, because the oval must be rotated 90° between passes (alternatively the rolls may be

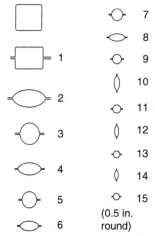

Figure 17.2 A bar rolling sequence dominated by oval/round passes. From R. Stewartson, *Metallurgical Reviews*, 4(16) (1959) 314. Copyright held by Maney Publishing, London, UK, www.maney.co.uk/journals/mr and www.ingentaconnect.com/content/maney/mtlr.

reoriented from stand to stand). The value of A_0 is $(\pi/4)Bb$, and h_0 is A_0/W_0, or $(\pi/4)B$. This second pass produces a round diameter of d, where W_1 is d, A_1 is $(\pi/4)d^2$, and h_1 is A_1/W_1 or $(\pi/4)d$. For simplicity, the ratio A_1/A_0 can still be $(1 - r)$.

17.2.2 Designing a pass sequence

The first step in designing a simple pass sequence is the designation of the starting and finishing round diameters, and the desired reduction per pass. Such a decision may reflect a wide variety of productivity and equipment capability considerations. For example, let us consider rolling from a diameter of 10 to 5 mm. The overall area reduction in terms of apparent true strain is the natural logarithm of $(10^2/5^2)$ or an apparent true strain of 1.386. Let us say that a ten-pass sequence (five oval/round sequences) is desired. The true strain of the reduction per pass would then be (1.386/10) or 0.1386. The corresponding decimal area reduction per pass is 0.1294, or 12.94%.

Now there is enough information to set up the pass sequence in a spread sheet format, *based on a given value of n_s*. Methods of determining n_s are discussed in Section 17.2.3. In any case, a spreadsheet "printout" is shown in Table 17.1. It can be seen that Pass 1 starts with a D value of 10 mm, and Pass 10 finishes with a d value of nearly 5 mm, as desired. The value of n_s selected for illustration purposes is 0.95. It can be seen that in Pass 1 the

Table 17.1 A ten-pass oval-round sequence in millimeters

Pass	D	B	b	d	r	n	A_0	w_0	h_0	A_1	w_1	h_1
1	10.00	10.37	8.40		0.1294	0.95	78.50	10.00	7.85	68.34	10.37	6.59
2		10.37	8.40	8.71	0.1294	0.95	68.34	8.40	8.14	59.50	8.71	6.83
3	8.71	9.03	7.31		0.1294	0.95	59.50	8.71	6.83	51.80	9.03	5.74
4		9.03	7.31	7.58	0.1294	0.95	51.80	7.31	7.08	45.10	7.58	5.95
5	7.58	7.86	6.37		0.1294	0.95	45.10	7.58	5.95	39.26	7.86	5.00
6		7.86	6.37	6.60	0.1294	0.95	39.26	6.37	6.17	34.18	6.60	5.18
7	6.60	6.84	5.54		0.1294	0.95	34.18	6.60	5.18	29.76	6.84	4.35
8		6.84	5.54	5.74	0.1294	0.95	29.76	5.54	5.37	25.91	5.74	4.51
9	5.74	5.96	4.82		0.1294	0.95	25.91	5.74	4.51	22.56	5.96	3.79
10		5.96	4.82	5.00	0.1294	0.95	22.56	4.82	4.67	19.64	5.00	3.93

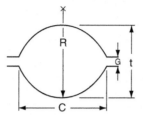

Figure 17.3 A simplified roll gap design system.

10 mm round is rolled to an oval of roughly 10.4 × 8.4 mm, with spread occurring from 10 to 10.4 mm. In Pass 2, the oval is rotated 90°, relative to the roll axis, with spread occurring from 8.4 to the 8.7 mm diameter of the round that will be going into Pass 3, and so on.

A simplified system for converting the values in Table 17.1 into groove and gap specifications is illustrated in Figure 17.3. To make specifications straightforward, a radius R is used, together with a groove width C, for both oval and round geometries. (The center from which R is drawn may or may not be within the area of the groove.) The relationship between R and C can be obtained from

$$C^2 = 4R(t - G) - (t - G)^2. \tag{17.3}$$

In the case of a pass designed to make a round of diameter d, the value of R will be d/2, the value of t will be d, and the value of C will be $(d^2 - G^2)^{1/2}$.

In the case of an oval, C is often chosen to be about ten percent larger than B, the intended major diameter. Having chosen C, R can be calculated from Equation 17.3, given that t must be b, the minor diameter, and given a desired value for G.

The value of G may depend on many factors, but practical values may be in the range of ten percent of the diameter, or average diameter, of the rod produced by the pass. When a single roll stand is used, a given value of G may be maintained for a number of passes. In principle, G reflects the starting roll gap plus the elastic separation of the rolls that occurs from the force developed during rolling. Since the rod diameter is generally a small fraction of the roll length, this elastic effect may be much less than experienced with the rolling of sheet and strip.

17.2.3 Analyzing existing passes and deducing the value of n_s

Let us suppose that the pass schedule set forth in Table 17.1 has actually been in use as an established technology. It might be useful to deduce the value of n_s for use in pass schedule redesign or pass geometry troubleshooting. The first step in such an effort would be careful measurement of the rod dimensions at the various stages of reduction. This would provide values for D, d, B, and b. From these data, values of A_0 and A_1 can be calculated and values of r determined. Taking the first pass as an example, it would be noted that W_0 or D was 10 mm, A_0 was 78.5 mm^2, W_1 or B was 10.37 mm, and A_1 was 68.34 mm^2. Substituting these values into Equation 17.2 allows calculation of an n_s value of 0.95. This process can be repeated for all ten passes, resulting in the same value of n_s, since Table 17.1 is not raw data, but a data set literally derived from an n_s value of 0.95. In practice, an existing satisfactory pass schedule will show some variation in back-calculated n_s, reflecting measurement scatter, minor wear patterns, an evolution in rod metallurgy, friction conditions, line tension, temperature, and so on. Unless a clear pattern of n_s-value shift is evident, process design and modification should utilize the average of the observed n_s values. Any passes manifesting truly anomalous n_s values are likely in need of correction, and may well be relatable to rod quality problems.

Further process design, and even the initial, characterizing rod measurements, should be undertaken in conjunction with determination of the values of C and G, as shown in Figure 17.3.

For cases where little exists in the way of current or well-documented past practice, n_s must be estimated from trials run on existing bar rolls. It may be instructive to vary the values of G in such trials to provide a wider range of rolling conditions, and to assess the sensitivity of n_s to given variations of G. Ideally, the roll pass geometry and general mill operation should be similar to that utilized in the system that is designed or corrected. If the only available mill is dissimilar, it should be understood that n_s is sensitive to pass geometry.

17.3. OTHER GEOMETRICAL SEQUENCES

Numerous sequences of geometry exist in rod, including diamond/ diamond, diamond/square, round-cornered-square, square/oval, and so on. For example, the pass necessary to convert profile 1 into profile 2 in Figure 17.2 is of the square/oval type. In all cases, the rod or rolls are rotated so that W_0 is less than W_1. All of these symmetrical shape sequences can be analyzed using the methodology of the type outlined in this chapter, although calculations must address corner radii, angles other than 90°, and so on.[126] Pass schedules utilizing asymmetrical shapes are beyond the scope of this chapter.

17.4. SOME PROCESS CORRECTION CONTEXTS
17.4.1 Appearance of "fins" and related "cold shuts"

When spread is not accommodated by pass geometry, metal is squeezed between the rolls, producing "fins" or "flash." The fins may break off, appearing as loose fines or as particles rolled into the rod surface. When they do not break off, they are rolled into the rod surface as "cold shuts," entrapping lubricant and oxide, compromising fatigue strength, compromising coatings, and so forth. Such flaws may be difficult to detect without expedients such as metallography or bend or twist testing. Figure 17.4 displays a cold shut revealed by metallography.

The appearance of fins in a previously satisfactory rolling system should be traceable to a given pass, or pair of passes, and it is necessary to sample stock between passes to locate the problem. Conditions leading to fins include the following:

Oversize W_0. Wear of the previous pass has resulted in an oversize W_0 for the next pass, leaving little room for spread.

Altered G value. The roll gap, G, may have been altered, either in the pass producing fins or in the preceding pass, perhaps in an effort to correct for wear.

Design or assembly error. An error may simply have been made in making the rolls or setting up the line.

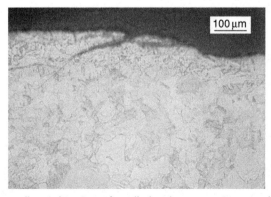

Figure 17.4 A metallographic view of a rolled rod cross section revealing a cold shut.

Once the location(s) of fin development has been determined, a comparison of the errant pass geometry with the original pass geometry, using the methodology outlined in this chapter, will generally indicate the solution to the problem. In some cases it will be clear that a revision is needed in the original roll pass design, particularly if rod quality and rolling conditions have evolved over the years.

17.4.2 General wear

For research and development evaluations, wear may be of little concern, and trial roll passes may even be undertaken on rolls that are not fully hardened. However, roll pass design pertinent to commercial production should anticipate roll wear. It can even be argued that, ideally, roll pass design geometry should reflect a mid-life state of wear. To pursue this idea, one must know the wear pattern that will result for a given initial pass geometry.

Apart from adjusting initial roll groove geometry in anticipation of wear, the only parameter available for pass geometry adjustment is the roll gap, G. Obviously G can be reduced as wear enlarges the groove. The consequences of this "correction" may not be fully appreciated, however, without subjecting the roll pass to analysis, for variations in G, at several stages of wear.

17.5. QUESTIONS AND PROBLEMS

17.5.1 Consider the oval at the end of second pass in Figure 17.2. Let the dimensions of the entering rectangle be 55 × 75 mm, and let the values of B and b for the exiting oval be 85 and 45 mm, respectively. What is the value of n_s?

Answer: Using Section 17.1.2, the values of A_0 and A_1, respectively, can be calculated as 4125 mm^2 and 3004 mm^2. Now, W_0 is 75 mm and W_1 is 85 mm. Putting these values into Equation 17.2, an n_s value of about 0.91 is obtained.

17.5.2 For this same pass, it is decided that a good value for R (see Figure 17.3) would be 60 mm. If G is only 1 mm, what will be the value of C? Is this value reasonable?

Answer: Using Equation 17.3 and recognizing that the value of t must be the same as b, or 45 mm, C can be calculated to be about 93 mm. This is reasonable, given that the value of B is only 85 mm.

CHAPTER EIGHTEEN

Wire Forming

Contents

18.1. SCOPE

Any operation that plastically deforms the wire *after* final wire drawing is considered wire forming for the purposes of this chapter. In most cases, this forming will be imposed on the wire by tool geometry. Manufacturing unit operations and tooling for wire forming are complex and varied, and this book does not attempt to review this technology in detail. On the other hand, the related end results of wire shape and cross section changes are excellent perspectives from which to consider the rudiments of wire forming, and this chapter will focus on these elements. The strain, stress, and force or moment implications of basic forming will be considered, including the role of elastic springback, once the tool is removed.

The basic forming modes to be considered are bending, twisting, stretching, upsetting, swaging, and cold extrusion.

18.2. BENDING

Figure 18.1 illustrates the geometry basic to the plane of symmetry of an initially straight wire section subject to bending. The innermost locus (top of the figure) of the wire is compressed in the longitudinal direction,

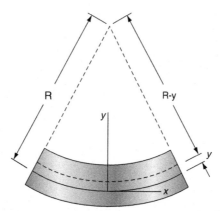

Figure 18.1 Bending geometry, longitudinal vertical section, plane of symmetry.

and the outermost locus (bottom of the figure) is in tension, in the longitudinal direction. There is a locus near the center of the wire, called the
neutral axis, along which the longitudinal strain is zero.

The longitudinal strain in the bent wire is

$$\varepsilon_x = -y/R, \tag{18.1}$$

where R is the bending radius, and y is the distance from the neutral
axis, where y is positive in the direction toward the center of the curvature.
If the neutral axis lies along the wire center, then the maximum absolute
value of the longitudinal strain is

$$|\varepsilon_x| = d/(2R), \tag{18.2}$$

where d is the wire diameter. At the inner locus of the bent wire the strain is
compressive at a value of $-d/(2R)$, and at the outer locus of the bent wire
the strain is $+d/(2R)$.

Now the neutral axis may not lie precisely along the wire center, and
when R is small, the inner and outer surfaces of the bent wire may be
complicated by superficial "buckling" or "fluting" instabilities on the compressive side, and by superficial "necking" or "flats" on the tensile side. Bending
may be undertaken to the point of surface cracking. However, for the most
general analysis of less severe bending, Equation 18.2 is quite instructive.

In the absence of residual stress, the wire will not undergo plastic bending
until the strain given in Equation 18.2 reaches the yield strain, ε_y, where:

$$\varepsilon_y = \sigma_y/E, \tag{18.3}$$

where E is Young's modulus. If the neutral axis lies along the wire center, the bending radius associated with this point of yielding is

$$R_y = dE/(2\sigma_y). \qquad (18.4)$$

At this point the bending moment borne by the wire cross section is

$$M_y = (\pi/32)\,d^3\sigma_y. \qquad (18.5)$$

As the bending radius becomes smaller, the wire cross section becomes progressively subjected to plastic flow, and, ignoring work hardening, the bending moment approaches a limiting value of

$$M_L = (0.167)\,d^3\sigma_y. \qquad (18.6)$$

This bending moment is approached rather rapidly, as the bending radius becomes smaller, and is a reasonable estimate of the actual bending moment for values of R as large as $(dE)/(4\sigma_y)$. Then the degree of springback to be expected once the wire is released from tool contact can be readily expressed, based on the removal of M_L. The resulting springback can be expressed as

$$1/R_1 = 1/R - (3.4)(\sigma_y/E)(1/d), \qquad (18.7)$$

where R_1 is the radius of curvature of the wire after springback.

As noted, this analysis assumes that the wire section is initially straight and does not include residual stress.

18.3. TWISTING

Figure 18.2 illustrates the geometry basic to the twisting of an initially straight and untwisted wire section. For analytical clarity, the wire section is

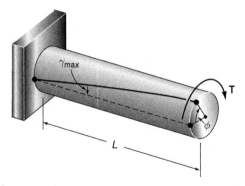

Figure 18.2 Twisting geometry.

mounted at one end. Along a length, L, the wire is subject to an angle of twist, ϕ. The engineering shear strain, γ, is

$$\gamma = r\phi/L, \qquad (18.8)$$

where r is the distance from the wire centerline. Thus, it is clear that the shear strain is zero at the wire center and maximum at the wire surface, and the maximum shear strain is

$$\gamma_{max} = d\phi(2L). \qquad (18.9)$$

In the absence of residual stress, the wire will not undergo plastic deformation until the strain given in Equation 18.9 reaches the yield strain in shear, γ_y, where

$$\gamma_y = \tau_y/G, \qquad (18.10)$$

and G is the shear modulus. The angle of twist associated with initial yielding is

$$\phi_y = (2L\tau_y)/(dG). \qquad (18.11)$$

At this point the torsional moment borne by the wire cross section is

$$T_y = (\pi/16)d^3\tau_y. \qquad (18.12)$$

As the twist angle becomes larger, the wire cross section becomes progressively subjected to plastic flow, and, ignoring work hardening, the torsional moment approaches a limiting value of:

$$T_L = (\pi/12)d^3\tau_y. \qquad (18.13)$$

This torsional moment is approached rather rapidly, as the twist angle becomes larger, and is a reasonable estimate of the actual torsional moment for values of ϕ as small as 2 ϕ_y. Now the degree of torsional springback to be expected once the wire is released from tool contact can be readily expressed, based on the removal of T_L. The resulting springback, ϕ_{sb}, can be estimated as:

$$\phi_{sb} = (8L\tau_y)/(3dG), \qquad (18.14)$$

for values of ϕ as small as 2 ϕ_y. For torsional moments just above yielding, values of ϕ_{sb} can be estimated from Equation 18.11.

As noted, this analysis assumes that the wire section is initially straight and does not include residual stress.

18.4. STRETCHING

Stretching in wire forming may be successfully undertaken to the extent of uniform elongation and no further. Once the necking instability develops, practical control of stretching is not possible. Uniform elongation is directly revealed by a tensile test, as discussed in Section 11.2. It is the strain at the peak of the engineering stress-engineering strain curve, or the strain at the ultimate tensile strength.

If the true stress-true strain curve is described the relation $\sigma = k\varepsilon^N$, then the true strain at necking, or the true uniform strain, has the value N.

The uniform elongation can be quite small in cold drawn rod or wire, since the conditions for necking are met early in the tensile test. This may be the case even when the rod retains considerable drawability. Hence, cold drawn wire is rarely suitable for stretch forming.

18.5. UPSETTING

For the purposes of this text, upsetting is defined as compressive plastic deformation, on a section of rod or wire, in the axial direction. Compression testing, as detailed in Sections 11.6 and 12.2.6, is upsetting from a deformation processing point of view, and the effects of friction and the limits of buckling and fracture are dealt with in these sections.

It is particularly noteworthy that axial upsetting deformation is reversed from drawing, and strength may be different from that displayed in drawing. The reversal of strain will often result in a decrease in flow stress, or even in work softening. Thus, there is considerable vulnerability to plastic buckling in upsetting of longer wire sections, especially sections with a length-to-diameter

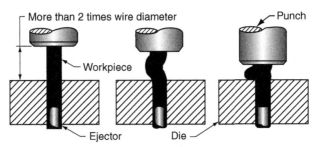

Figure 18.3 Schematic illustration of buckling during heading. From Forming, Cold Heading and Cold Extrusion, R. N. Wright (Ed.), *Metals Handbook Desk Edition*, American Society for Metals, Metals Park, OH, 1985, 26–48. Copyright held by ASM International, Materials Park, OH.

ratio of two or more. Figure 18.3 illustrates the development of buckling in a heading (upsetting) operation.[128]

18.6. SWAGING

18.6.1 First case

For the purposes of this text, swaging is defined as compressive plastic deformation, on a section of wire, perpendicular to the axial direction. In one class of swaging, the tool length along the wire axis is many times the rod or wire diameter. This process is sometimes referred to as <u>sidepressing</u>. Sidepressing involves little increase in the wire length, and plane strain analysis is relevant. A schematic description of a sidepressed cross section is given in Figure 18.4.[129] For low friction conditions, the shape change can be described by the relation

$$x = [1/(4h)][A_0 - R_0^2(\theta - \sin\theta)], \tag{18.15}$$

where x is the width of contact with the die, h is the half-height of the cross section perpendicular to the die face, A_0 is the wire cross-sectional area, R_0 is the initial wire radius, and θ is $2 \sin^{-1}(h/R_0)$. The relationship of

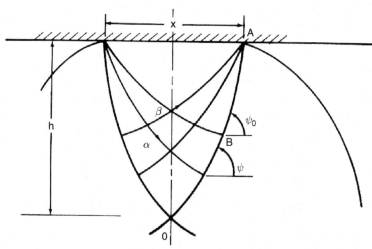

Figure 18.4 Sidepressing deformation geometry with slip line field. From A. T. Male and G. E. Dieter, *Hot Compression Testing, Workability Testing Techniques*, G. E. Dieter (Ed.), American Society for Metals, Metals Park, OH, 1984, 70. Copyright held by ASM International, Materials Park, OH.

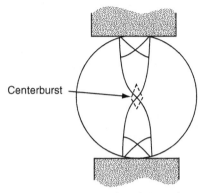

Centerburst

Figure 18.5 Sidepressing geometry with slip line field and centerburst.
From W. A. Backofen, *Deformation Processing*, Addison-Wesley Publishing Company,
Reading, MA, 1972, 148. Copyright held by Pearson Education, Upper Saddle River, NJ, USA.

the width of the sidepressed cross section, w, to h can be approximated with
the expression

$$(w/2)^2[\pi - \cos^{-1}(h/w)] = A_0. \tag{18.16}$$

Sidepressing imposes a high degree of tensile stress at the rod or wire
centerline. This poses a major risk of centerline fracture and, in fact, sidepressing is often used as the basis of workability testing, since fracture
conditions may be developed at the centerline *and* at the lateral surfaces.
Plane strain slip line field analysis projects a centerline hydrostatic tension
level of $(0.63)\sigma_o$ upon initiation of sidepressing. This is comparable to
hydrostatic stress values in the neck of a tensile specimen or at the centerline
when drawing wire with a Δ value of 6 or 7. Figure 18.5 illustrates
development of a centerline fracture with plane strain sidepressing.[130]

18.6.2 Second case

The tool length along the wire axis is the same order as the rod or wire
diameter in a second case of swaging. There is also significant increase in rod
length, and plane strain conditions are not met. This case is analogous to that
of *cogging* in forging practice, where a limited amount of the workpiece is
upset in each operation, with the upset target progressively moved along the
length of the workpiece, from blow to blow.

A schematic representation of cogging is presented in Figure 18.6. The
principal concern is the amount of spread, or increase in w, as compared to

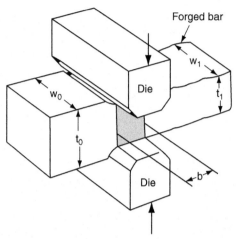

Figure 18.6 Cogging geometry. From A. Tomlinson and J. D. Stringer, *Journal of the Iron and Steel Institute*, London, 193 (1969) 157.

the decrease in thickness, or reduction in t. Tomlinson and Stringer[131] have addressed this concern with a coefficient of spread, S_c, where

$$S_c = (w_1 - w_0)/(t_0 - t_1), \tag{18.17}$$

where w_0 and w_1 are the initial and final widths, respectively, and t_0 and t_1 are the initial and final thicknesses, respectively. Using Δ and friction hill arguments, S_c should be a function of $(t/w)/(t/b)$, or a function of (b/w), where b is the tool dimension in the longitudinal direction of the work-piece. Consistent with this, Tomlinson and Stringer found that

$$S_c = (b/w_0)/[1 + (b/w_0)]. \tag{18.18}$$

When applying Equation 18.18 to wire swaging, it should be noted that the value of b in cogging is almost the same as the contact length of the swaging die along the length of the wire. The value of w_0 in cogging is larger than the *average* contact length, w_{sw}, of the swaging die with the wire in the direction perpendicular to the wire axis. If w_{sw} is less than w_0, it follows that

$$S_c = (b/w_0)/[1 + (b/w_0)] < (b/w_{sw})/[1 + (b/w_{sw})] = S_{sw}, \tag{18.19}$$

where S_{sw} is the coefficient of spread in round wire swaging. The value of w_{sw} will increase with the degree of strain in swaging, whereas b will not change. Hence, S_{sw} will generally decrease as the degree of strain in swaging increases.

There are many additional factors to consider regarding spread in swaging, such as surface quality, friction, workpiece texture, and so forth. Hence, it is often better to use Equation 18.19 only as a rough process design guideline. In this regard, it may be useful to replace w_{sw} with $d/2$, where d is the wire diameter. With this approximation,

$$S_{sw} = (2b/d)/[1 + (2b/d)] = (w_1 - d)/(d - t_1), \qquad (18.20)$$

where w_1 and t_1 are the width and thickness of the as-swaged section in direct analogy to their role in Equation 18.17.

18.6.3 Rotary swaging

Rotary swaging is a process where the "second case" swaging outlined in Section 18.6.2 is repeatedly applied, as the swaging direction is rotated about the rod or wire axis. During this repeated rotary swaging, the rod or wire is advanced into the swaging die in partial analogy to cogging. Figure 18.7 displays an operator advancing a rod into a swaging machine, and Figure 18.8[132] presents a schematic drawing of the reciprocal action imposed on the swaging dies by a roll and ring assembly.

The cumulative effect of rotary swaging involves no spreading and only elongation, somewhat analogous to rod and wire drawing. Rotary swaging can be applied continuously over long lengths of wire and rod, but the individual swaging blows are somewhat shallow, a twist is developed, and

Figure 18.7 An operator advancing a rod into a swaging machine.

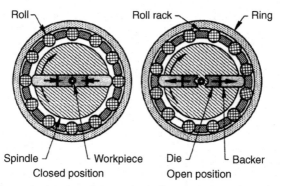

Figure 18.8 Schematic drawing of the reciprocal action imposed on a swaging die by a roll and ring assembly.
From Forming, Forming Bars, Tube and Wire, R. N. Wright (Ed.), *Metals Handbook Desk Edition*, American Society for Metals, Metals Park, OH, 1985 26–35. Copyright held by ASM International, Materials Park, OH.

much redundant work is produced. Also, the surface is much less smooth than that of a drawn workpiece. Still, rotary swaging is quite useful for pointing, tapering, and sizing.

 18.7. COLD EXTRUSION

Extrusion involves the pushing of a workpiece of a given cross-sectional area into a die or tool resulting in an extrudate of reduced cross section. In many ways the process is like drawing, except that the workpiece is pushed into, rather than pulled though, the die. The dominant extrusion technology is hot extrusion, particularly of aluminum and aluminum alloys. However, short lengths of rod and wire are often extruded without prior heating using a technology called cold extrusion. Such processing may also be referred to as cold forging, and the two topics are often discussed together. This section will only address cold extrusion.

A simple expression for extrusion pressure, P, is

$$P = k\,\sigma_o \ln R, \qquad (18.21)$$

where R, the *extrusion ratio*, is the ratio of starting cross-sectional area to final cross-sectional area and where k is a coefficient reflecting friction and redundant work. Figure 18.9 displays pressure versus tensile strength data for a *forward* cold extrusion, with implicit k values of three to four. Figure

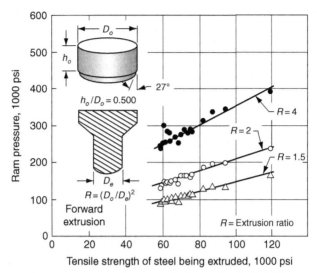

Figure 18.9 Pressure versus tensile strength for a forward cold extrusion. From Forming, Forming Bars, Tube and Wire, R. N. Wright (Ed.), *Metals Handbook Desk Edition*, American Society for Metals, Metals Park, OH, 1985 26–50. Copyright held by ASM International, Materials Park, OH.

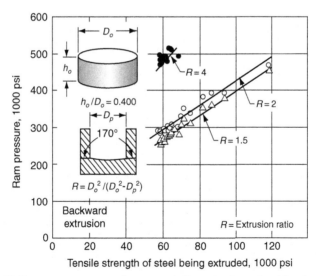

Figure 18.10 Pressure versus tensile strength for a backward cold extrusion. From Forming, Forming Bars, Tube and Wire, R. N. Wright (Ed.), *Metals Handbook Desk Edition*, American Society for Metals, Metals Park, OH, 1985 26–50. Copyright held by ASM International, Materials Park, OH.

18.10 displays pressure versus tensile strength data for a *backward* cold extrusion, with implicit k values of six to ten.[133]

A more elaborate model of extrusion pressure is

$$P = 2\tau_o[(mA_{sc}/2A) + (\Phi \ln R)], \qquad (18.22)$$

where m is the friction factor, A_{sc} is the area of sliding contact, A is the initial cross-sectional area, and Φ is the redundant strain factor. This model is comparable to the semi-empirical model:

$$P = 2\tau_o(a + b \ln R), \qquad (18.23)$$

where "a" generally reflects friction and "b" reflects redundant and non-uniform flow.

 ## 18.8. QUESTIONS AND PROBLEMS

18.8.1 A 1 cm diameter rod with a tensile yield strength of 200 MPa is bent until nearly fully plastic. What is the bending moment? The same rod is twisted until nearly fully plastic. What is the torsional moment?
Answer: Using Equation 18.6, the bending moment is readily calculated as 33.4 Nm. Using Equation 18.13 and approximating the torsional (shear) yield strength as one-half of the tensile yield strength, the torsional moment is calculated to be 26.2 Nm.

18.8.2 A 1 cm diameter rod is sidepressed from a thickness of 1 cm to a thickness of 0.95 cm. What is the width of contact with the die?
Answer: Use Equation 18.15 to calculate the width of contact, x where h is 0.475 cm, A_0 is 0.785 cm^2, R_0 is 0.5 cm, and θ is 2.5 radians. Thus, the width of contact is 0.324 cm.

18.8.3 A 1 cm diameter rod is swaged under conditions where S_{sw} is 0.8. If the thickness is reduced to 0.95 cm, what is the width of the swaged section?
Answer: Using Equation 18.20, w_1 may be calculated to be 1.04 cm.

18.8.4 Using Equation 18.21, determine the k values implicit in Figure 18.9.
Answer: Approximating the flow stress with the tensile strength, and using a value of 80 ksi, the respective pressures for R values of 1.5, 2, and 4 are about 120, 180, and 300 ksi. On this basis the respective k values are 3.7, 3.2, and 2.7. Clearly the k values decline as the R values increase. On the other hand, a k value of 3.5 would be useful for rough estimates at light reductions.

CHAPTER NINETEEN

Physical Properties

Contents

19.1. SCOPE

This chapter reviews the physical properties that appear in the various formulas and figures in the text, specifically density, melting temperature and range, specific heat, thermal conductivity, and electrical resistivity.

19.2. DENSITY

Density is the mass of a material per unit volume of that material. The most common units in modern technical literature are grams per cubic centimeter. Some data are presented in units of pounds (mass) per cubic foot. Density values in pounds (mass) per cubic foot can be multiplied by 1.60×10^{-2} to obtain the numerical value for grams per cubic centimeter.

Theoretical densities of pure crystalline substances can be calculated from the unit cell of the crystal structure. For example, Figure 11.10 shows a face-centered-cubic (FCC) structure, and the theoretical density of copper (an FCC metal) can be calculated, for ambient temperature, based on the atomic radius and atomic weight of copper.

Generally speaking, handbook density values reflect physical measurements and not theoretical calculations. However, these measurements reflect variations due to porosity, inclusions, and experimental uncertainties. Under the best of circumstances, density measurements are only accurate to four significant figures.

Wire Technology
ISBN 978-0-12-382092-1, DOI: 10.1016/B978-0-12-382092-1.00019-1

Approximate ambient temperature densities for some common metals are listed in the following table.

Material	Density (g/cc)	Material	Density (g/cc)
Aluminum	2.71	Nickel	8.89
Brass (cartridge)	8.53	Platinum	21.45
Carbon steel	7.85	Silver	10.49
Copper	8.89	Stainless steel(304)	8.00
Gold	19.32	Tantalum	16.6
Iron	7.87	Titanium	4.51
Molybdenum	10.22	Tungsten	19.3

19.3. MELTING POINTS AND RANGES

Melting temperatures are important for numerous reasons. First, it is often necessary to consider the potential for melting of a wire material in a service aberration (fire, catastrophic friction, etc.). Second, the melting temperatures say much about the primary processing technology. Most important, properties at ambient temperatures and service temperatures, in general, will often be a function of the *homologous temperature*, T_H, where

$$T_H = T/T_M, \qquad (19.1)$$

and where T is the absolute service temperature and T_M is the absolute melting temperature. For example, lower homologous temperatures are associated with lower elastic moduli and higher coefficients of thermal expansion.

Pure metals, as opposed to alloys, have specific melting temperatures. Most alloys, however, have a range of temperatures where both liquid and solid phases are present. This "mushy" temperature range is bounded at the low end by the *solidus* temperature, T_S, and on the high end by the *liquidus* temperature, T_L. For practical purposes, the solidus temperature will be considered as the effective melting temperature for such alloys.

Melting temperatures, T_M, and ambient temperature homologous temperatures, T_H, for some common metals are listed in the following table.

Material	T_M, T_S (°C)	T_H	Material	T_M, T_S (°C)	T_H
Aluminum	660.4	0.314	Nickel	1455	0.170
Brass (cartridge)	925	0.245	Platinum	1772	0.143
Carbon steel	1493	0.166	Silver	962	0.237

Copper	1085	0.216	Stainless steel(304)	1400	0.175
Gold	1064	0.219	Tantalum	2996	0.090
Iron	1538	0.162	Titanium	1668	0.151
Molybdenum	2617	0.101	Tungsten	3410	0.080

 ## 19.4. SPECIFIC HEAT

The specific heat of a material is the heat, or thermal energy, required to raise a given mass by one degree of temperature. The most common units in the modern technical literature are joules per kilogram per degrees Kelvin. Some data are presented in units of Btu per pound (mass) per °F. Heat capacity values in Btu per pound (mass) per °F can be multiplied by 4.18×10^3 to obtain the numerical value for joules per kilogram per degrees Kelvin.

Approximate specific heat values for some common metals at ambient temperature are listed in the following table.

Material	Specific Heat(J/kg-K)	Material	Specific Heat(J/kg-K)
Aluminum	904	Nickel	456
Brass (cartridge)	375	Platinum	132
Carbon steel	486	Silver	235
Copper	385	Stainless steel (304)	500
Gold	128	Tantalum	139
Iron	486	Titanium	528
Molybdenum	276	Tungsten	138

 ## 19.5. THERMAL CONDUCTIVITY

The thermal conductivity of a material is the constant relating the steady-state heat flow, in energy per unit area per unit time, to the temperature gradient in degrees per unit length. The most common units in modern technical literature are watts per meter per degrees Kelvin, where a watt is one Joule per second. Some data are presented in units of Btu per foot per hour per °F. Thermal conductivity data in Btu per foot per hour per °F can be multiplied by 1.728 to obtain the numerical value for watts per meter per degrees Kelvin.

With metals, thermal and electrical conductivity are closely related, since electron motion is the principal vehicle in each case.

Approximate thermal conductivity values for some common metals at ambient temperature are listed in the following table. Thermal conductivity in metals decreases substantially with increases in temperature.

Material	Thermal Conductivity (W/m-K)	Material	Thermal Conductivity (W/m-K)
Aluminum	222	Nickel	70
Brass (cartridge)	120	Platinum	71
Carbon steel	51.9	Silver	428
Copper	388	Stainless steel (304)	16.2
Gold	315	Tantalum	54.4
Iron	51.9	Titanium	16
Molybdenum	142	Tungsten	155

19.6. ELECTRICAL RESISTIVITY

The electrical resistivity of a length of wire is its electrical resistance, in ohms, times its cross-sectional area, divided by its length. Common basic units of electrical resistivity are ohm meters (Ω-m) or micro-ohm centimeters ($\mu\Omega$-cm). Electrical resistivity in ohm meters is multiplied by 10^8 to obtain the numerical value for micro-ohm centimeters.

Electrical conductivity is simply the inverse of electrical resistivity, with units of $(\Omega$-m$)^{-1}$, and so forth. As noted in Section 19.5, in the case of metals, electrical and thermal conductivity are closely related, since electron motion is the principal vehicle in each case.

Approximate electrical resistivity values for some common metals at ambient temperature are listed in the following table. Electrical resistivity in metals increases substantially with increases in temperature.

Material	Electrical Resistivity ($\mu\Omega$-cm)	Material	Electrical Resistivity ($\mu\Omega$-cm)
Aluminum	2.9	Nickel	9.5
Brass (cartridge)	6.2	Platinum	10.6
Carbon steel	16	Silver	1.47

Copper	1.72	Stainless steel (304)	72
Gold	2.35	Tantalum	13.5
Iron	6.2	Titanium	47
Molybdenum	5.2	Tungsten	5.3

CHAPTER TWENTY

Current and Near-Term Developments

Contents

20.1. SCOPE

This chapter presents a number of current and near-term developments that illustrate the advances and opportunities of the contemporary wire industry. A number of commercial examples have been identified, solely at the discretion of the author, for the purposes of this text. The author does not necessarily endorse the products, has not tested these products, and cannot explicitly confirm their reported capabilities. However, the products have been selected from high visibility venues, and from generally reliable sources. In particular, the author has relied upon the judgment of the editorial staff of *Wire & Cable Technology International*, in their determination of awards for Top Products of 2008 and Top Products of 2009.[134,135] Each of the following examples has received a Top Products award.

Wire Technology
ISBN 978-0-12-382092-1, DOI: 10.1016/B978-0-12-382092-1.00020-8

20.2. DRAWING MACHINES

20.2.1 Zero-slip rod breakdown machine

As discussed in Section 9.6.6, many drawing systems involve slip of the wire on the capstans. Designing for this slip allows the easy accommodation of changes in wire velocity due to die size variations and die wear. Some degree of capstan wear can be accommodated; however, this slip promotes capstan wear and compromises wire surface quality. The issue of wire surface quality is extremely important to magnet wire processing, as discussed in Section 8.8.3 and illustrated in Figure 8.21.

In this context, FRIGECO, a division of Mario Frigerio of Italy has developed a TA model *rod breakdown* machine "with absolute zero slip for the efficient production of superior quality copper and aluminum wires." Photographs of the machine are presented in Figures 20.1 and 20.2. In the words of the vendor:

On the TA model machine, each drawing block is driven by an independent AC motor with the synchronization wire-capstan speed performed by a dedicated dancer. This configuration eliminates any kind of longitudinal slipping problem to avoid longitudinal wire scratches. In each drawing block, the wire is wound to a double capstan. The first pulling capstan is motor driven, while the second idle capstan is slightly out of center with respect to the first. The second idle capstan allows spacing of the wraps in order to prevent side sliding and drafting to exclude any axial wire scratches. The machine is also supplied with large diameter

Figure 20.1 FRIGECO TA "zero-slip" rod breakdown machine. (Courtesy of Initial Publications, Inc., Akron, OH, USA).

Figure 20.2 TA "zero-slip" machine features independently driven drawing blocks and wire wound onto double capstans. (Courtesy of Initial Publications, Inc., Akron, OH, USA).

pulling and idle capstans to prevent dangerous wire yielding. Additionally, a high lubricant pressure die holder and efficient wire cooling help in achieving high production speeds.

20.2.2 Electronically driven intermediate wire drawing machine

An *intermediate* wire drawing machine designed for high-surface quality non–ferrous drawing is offered by Maschinenfabrik Niehoff GmbH & Co. KG of Schwabach, Germany. This electronically driven MSM224 drawing machine is capable of drawing wires with inlet diameters of 3.5 mm to a final diameter range of 0.2 to 1.8 mm, with line speeds up to 25 m/s. A photograph of the machine is presented in Figure 20.3. In the words of the vendor:

The drawing capstans are individually drawn by water-cooled AC motors. The drive principle enables a minimized slip operation resulting in wire with a very high surface quality. Other characteristics include a completely submerged drawing section and die holders with high pressure lubrication. For soft wire production, the MSM224 is designed to operate in-line with an inductive continuous annealer....

Figure 20.3 Maschinenfabrik Niehoff MSM224 electronically driven intermediate drawing machine. (Courtesy of Initial Publications, Inc., Akron, OH, USA).

20.3. LONG-LIFE DIES FOR FERROUS DRAWING

Long die life can be elusive in many ferrous drawing operations. Diamond dies, as noted in Section 9.2.5, appear to react with carbide-forming elements such as iron, thus limiting their usefulness in steel drawing applications. Tungsten carbide–cobalt dies function satisfactorily in the drawing of steel, but die life is relatively short.

Therefore, there is a niche in the marketplace for a long-life die that is compatible with steel drawing. Such a product has been introduced by Allomet Corporation, North Huntingdon, Pennsylvania. The die material is called Tough Coated Hard Powder (TCHP). An SEM micrograph of the structure of TCHP is shown in Figure 20.4. The dark circular areas in the microstructure are hard alumina (Al_2O_3) core particles. The core particles are surrounded by a WC shell, and the extended bright regions represent a tungsten carbide (WC) and cobalt (Co) matrix. The WC shell thickness is 100–200 nm, and the matrix is comprised of 100–500 nm WC crystals in Co. Alternative formulations involve core particles of diamond, cubic boron nitride, titanium carbo-nitride, and so forth.

Field trials and analyses of TCHP performance in steel drawing have revealed a seven-fold increase of TCHP die life over that of WC-Co dies.

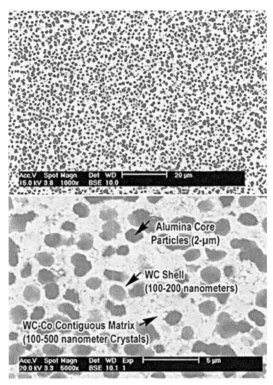

Figure 20.4 SEM micrograph of TCHP die material structure. (Courtesy of Initial Publications, Inc., Akron, OH, USA).

20.4. MEASUREMENT AND INSTRUMENTATION

20.4.1 Diameter measurement systems

The importance of wire diameter and cross section shape to die wear monitoring, die alignment, and so forth has been discussed in Section 9.5. Many wire producers depend on static, individual-location measurements; however, modern metrology and real-time data analysis have much to offer the wire industry, and new and upgraded products continue to appear. LaserLinc, Inc. of Yellow Springs, Ohio, offers a versatile non-contact ultrasonic system for wire diameter measurement. The system is called UltraGauge + and it is suitable for in-process tolerance checking, trending, SPC, feedback control, data logging, and "recipes." A photograph of the digital signal processor unit is shown in Figure 20.5.

Another diameter measurement system is that presented by ODAC® diameter scanners, with calibrated single scan (CSC) technology. This

Figure 20.5 Digital signal processor unit for UltraGauge$^+$ ultrasonic measurement system. (Courtesy of Initial Publications, Inc., Akron, OH, USA).

Figure 20.6 ODAC® heads with calibrated single-scan technology for diameter measurement. (Courtesy of Initial Publications, Inc., Akron, OH, USA).

equipment is offered by Zumbach Electronics Corporation, Mt. Kisco, New York. A photograph of the diameter measurement system is shown in Figure 20.6 In the words of the vendor:

> *With rates of 2400 scans per second and more, each scan is identified and calibrated for best accuracy. This technology can reduce foam shrinkage in extrusion. A dedicated processor analyzes the integrity of each scan relative to product surface, dirt, water droplets, etc., and compiles the measurements to provide minimum, maximum and average diameter, with ovality and flaw detection.*

20.4.2 Fault detector

The discussion of wire breaks in Chapter 12 makes it clear that early, real-time warning of flaw development can help avoid lost production time in restringing. Moreover, fault detection systems have continued to expand

Figure 20.7 Components of the Roland Electronic UFD40 fault detection system. (Courtesy of Initial Publications, Inc., Akron, OH, USA).

in capability. Integration of detector data with process control is offered with a UFD40 system available from Roland Electronic GmbH of Keltern, Germany. A photograph of the fault detector components is shown in Figure 20.7. As noted by the vendor:

> *The UFD40 has all the classic features of eddy current testing equipment used in automated inspection lines such as variable operating frequencies, high pass and low pass filters, y-component and vector analysis. This makes it possible to configure the same hardware specifically to the inspection task under consideration. Measurement is based on the eddy current principle whereby the wires and cables pass through encircling coils sensors in the continuous production process. The sample rate of measurement is 15 kHz. This means materials passing the sensors at 10 mps result in a resolution of 0.6 mm in length.*

20.5. ANNEALING

In-line or tandem annealing is widely used in non–ferrous drawing, as noted in Section 13.4, and it is hoped that the technology can be increasingly applied to ferrous drawing. While conventional in-line annealing has been highly engineered, contact of the wire with the sheaves can present problems of arcing and general surface damage. Hence, development of "contactlesss" annealing technology has been and continues to be a worthy objective.

Figure 20.8 NIEHOFF-BÜHLER contactless annealing system. (Courtesy of Initial Publications, Inc., Akron, OH, USA).

In this context, a contactless annealing and preheating system has been developed by NIEHOFF-BÜHLER GmbH NBM of Schwabach, Germany. A photograph of the contactless system is shown in Figure 20.8. The vendor describes the system as follows:

> *The MT200/R1170.2VA data and telephone wire drawing line incorporates annealing and preheating technology to provide a system designed for in-line production of wires for data and telephone cables. The technology that is incorporated in this wire drawing line was derived from the inductive annealers of the joint venture company…. The simple and contactless annealing principle generates a high wire surface quality, while also achieving a considerable reduction in the operational costs through the omission of carbon brushes.*

 20.6. PAYOFF AND TAKE-UP SYSTEMS

Much of the capital cost and materials handling efficiency of a wire drawing system lies in payoff and take-up equipment. This has become increasingly the case as drawing speeds and productivity expectations have increased. Hence, new equipment in this area should be given careful scrutiny.

20.6.1 Flyer arm payoff

Wyrepak Industries, Inc., Middletown, Connecticut, offers a BMPFT flyer arm unit as an improved payoff for multi-wire, single strand bare wire and

Figure 20.9 Wyrepak Industries flyer arm payoff. (Courtesy of Initial Publications, Inc., Akron, OH, USA).

insulated copper and steel wire. A photograph of the payoff system is shown in Figure 20.9. The vendor remarks that:

> Features of the BMPFT Flyer Arm payoff include a full depth leading traverse roller designed to take the wire out (and not up), adjustable arm length (optional), fixed or swivel top exit pulley, dynamically balancing and the capability to handle reel sizes 12" to 49" (DIN 300 to 1250). Installation of the BMPFT Flyer Arm unit is simple with bore mounted clamping adapter.

20.6.2 Micro-wire payoff and feeder

The challenges of payoff and take-up increase as one produces fine wire. The losses and costs involved with breaks are especially critical, as discussed in Section 12.4.4. In this context, a payoff and feeder system designed for fine wire is offered by the Kine-Spin/Sleeper Division — Kinefac Corporation of Worcester, Massachusetts. Called the Micro-Dereeler, the unit handles wire from 25 to 250 μm diameter in loose coils or on spools. A photograph of the product is shown in Figure 20.10. The vendor notes that:

> Driven by a variable speed DC drive, the intermittent wire feed is accommodated by a moving arm three loop dancer which can store 36" of wire. It is equipped with an adjustable precision counterbalance system that actuates the DC drive. The low mass anti-friction mounted storage rolls minimize startup drag. The feed

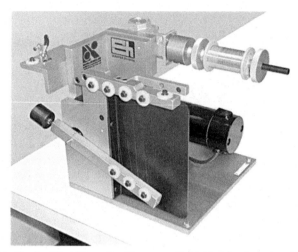

Figure 20.10   Kinefac Micro-Dereeler. (Courtesy of Initial Publications, Inc., Akron, OH, USA).

cycle is electronically tunable to be in phase with the machine being fed. The Micro-Dereeler provides superior control feed tension needed for repeatable high speed forming, coiling or cutoff output from micro machines. The spool support mounting shaft axis is horizontal, and the loose coil support axis is vertical to prevent inter-coil sagging. The loose coil support unit has a single point coil ID adjustment. The standard feed is from the left side of the unit and a set of reversal brackets and rollers are available for right hand or vertical feed.... The machine structure is of anodized aluminum and its function does not require any lubrication, therefore it is operable in a clean room environment.

20.6.3 Gantry system payoffs and take-ups

Flexibility and mobility are important considerations in many payoff and take-up situations. A gantry system offered by Tulsa Power, Inc., Tulsa, Oklahoma, appears to offer many conveniences in this regard. A photograph of the product is shown in Figure 20.11. The vendor describes the 100K Gantry Systems line as follows:

Reel capacities for these machines range from 2175 mm through 4350 mm flange diameter, reel widths to 4572 mm and maximum reel weights to 100,000 lb. The entire structure offers easy loading with true walk-through capabilities. The telescoping superstructures portal design conforms to the size of reel to minimize floor space requirements. Each system traverses on rails to offer material payout or takeup with no fleeting angle. Electromechanical drives are utilized to allow for up/down and in/out manipulation of the main frame during reel loading and un-loading sequences.

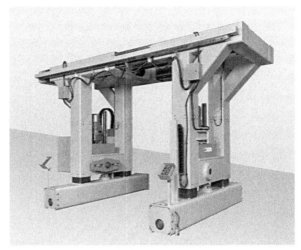

Figure 20.11 Tulsa Power gantry system. (Courtesy of Initial Publications, Inc., Akron, OH, USA).

20.6.4 High-speed spooler

As noted earlier, speed is often a major issue in wire processing, and there are usually new products introduced with increasing operating speeds or increased reliability at high speed. Thus, the High Speed SW6 spooler offered by Wyndak Inc., Hickory, North Carolina, is of interest. A photograph of the product is shown in Figure 20.12, and the unit is described as follows:

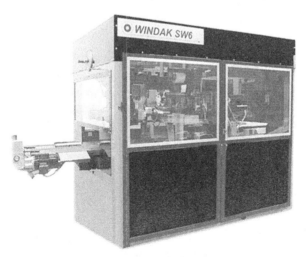

Figure 20.12 Wyndak high speed SW6 spooler. (Courtesy of Initial Publications, Inc., Akron, OH, USA).

...the SW6 has successfully achieved a rate of six 100 m spools produced per minute. The SW6 spooler is designed for a maximum output of six 100 m spools per minute, with a guarantee by the manufacturer of five 100 m spools per minute. The SW6 High Speed spooler's dual lift system allows for cycle times of less than 10 seconds. Stop time is less than one second for the optimized production of spools measuring 165 mm (6.5") OD × 90 to 200 mm (3.5" to 8") wide.

LIST OF SYMBOLS—ENGLISH ALPHABET

A_f	cross-sectional area at fracture
A_{fl}	cross-sectional area of a flaw
A_0	initial cross-sectional area
A_1	cross-sectional area after reduction, or length extension
A_n	cross-sectional area after n^{th} drawing pass
A_{sc}	area of sliding contact
AWG	American Wire Gage
b	ratio of back stress to average flow stress; minor diameter of an oval; swaging die dimension along wire axis
B	number of drawing breaks; major diameter of an oval
Btu	British thermal unit
B&S	Brown and Sharpe gage
BHN	Brinell hardness number
c	material property reflecting workability
C	specific heat
C_{Al}	concentration of aluminum
C_0	initial concentration
C_1	final concentration
d	wire, rod, or bar diameter
d_g	grain diameter
d_i	hardness indentation diameter
d_0	initial wire, rod, or bar diameter
d_1	diameter after reduction, or length extension
D	drawing block, capstan, mandrel, or entering round rolling stock diameter
D_c	corner-to-corner diameter
D_f	flat-to-flat diameter
D_i	hardness indenter diameter
E	Young's modulus
F	force
F_r	roll force
g	roll gap
G	strain rate strength coefficient; shear modulus
h	height of a compression test specimen; half-height of swaged cross section
h_0	height of rectangle of width W_0
h_1	height of rectangle of width W_1
H	hardness
I	annealing index
IACS	International Annealed Copper Standard
J	wire material property inversely related to toughness
k	strength coefficient in work hardening curve fit; extrusion coefficient
k_g	grain boundary strength coefficient
K	thermal conductivity

KHN	Knoop micro-hardness number
l_0	initial gage length
l_1	extended gage length
L	length
L_c	contact length of wire with die
L_d	length of deformation zone in the die
L_{eq}	length downstream from die where temperature equilibration occurs
L_r	length of roll–workpiece contact
$L_{sliding}$	distance of sliding in wear analysis
m	friction factor
M	mass
M_y	bending moment at yielding or plastic flow
M_G	net mass of alloy melt removed per unit wire surface area
M_{Go}	mass of alloy overlay that solidifies per unit wire surface area
M_L	maximum bending moment
n	pass number
n_s	spread index
N	number of wraps of the wire around a capstan; exponent in work hardening curve fit
N_{fl}	number of flaws per unit volume
N_t	number of twists
N_{tmax}	number of twists to failure
P	average die pressure; extrusion pressure
P_o	average die pressure in the absence of back tension
P_r	rolling power
q	proportionality constant in the Archard equation; material property reflecting surface workability
Q	activation energy
Q_d	die life constant equal to $H\delta/(2q)$
r	drawing or rolling reduction, either in decimal form or percent; distance along radial axis
r_n	reduction in the nth drawing pass
R	gas constant; bending radius; extrusion ratio
R_r	roll radius
R_s	side roll radius
R_y	bending radius at initial yielding
R_0	radius of curvature of mandrel; initial wire radius
R_1	radius of curvature after springback
s	speed
S_c	spread coefficient
S_{sw}	spread coefficient for round wire swaging
SI	International System measurement system
t	thickness of swaged cross section
t_{eq}	time required for equilibrium of frictional heating
t_{life}	die life
t_0	initial workpiece thickness
t_1	workpiece thickness after pass

T	temperature
T_{eq}	equilibrated temperature
T_f	volume-averaged drawing temperature increment from friction work
T_{max}	maximum value of drawing temperature increment
T_{fmax}	maximum value of drawing temperature increment from friction work
T_o	reference temperature
T_{rw}	drawing temperature increment from redundant work
T_{uw}	drawing temperature increment from uniform work
T_w	drawing temperature increment from total deformation work
T_y	torsional moment at yielding or plastic flow
T_H	homologous temperature
T_L	maximum torsional moment, liquidus temperature
T_M	melting temperature
T_S	solidus temperature
T_r	roll torque
T_0	temperature of wire prior to die entry
UK/US	United Kingdom and United States measurement system
UTS	ultimate tensile strength
v	drawing velocity or speed
V_0	velocity at the die entrance
V_1	velocity at the die exit
V_n	velocity at the die exit for the nth pass
$V_{capstan}$	capstan surface velocity
V_{wear}	volume of material worn away
w	work per unit volume; width of workpiece cross section
w_f	volume averaged friction work
w_r	volume averaged redundant or non-uniform work
w_{sw}	average width of swaged cross section
w_u	uniform work per unit volume
w_0	initial workpiece width
w_1	workpiece width after pass
W	work
W_0	Rod width before rolling pass
W_1	Rod width after rolling pass
X	width of die contact
y	distance from wire center, in the plane of bending, positive in direction toward center of curvature
y_o	lubricant film thickness at drawing zone entry
z	pressure coefficient

LIST OF SYMBOLS — GREEK ALPHABET

α	die semi-angle or half-angle; 2α is die included angle
α_{opt}	die semi-angle value for minimum Σ
β	die semi-angle or half-angle; 2β is die included angle
δ	average die diameter increase due to wear
Δ	drawing die deformation zone shape (average diameter/length)
Δ'	plane strain Δ
Δ_{ave}	average value of Δ
Δ'_{ave}	average value of plane strain Δ
Δ_{opt}	Δ value for minimum Σ
ε	normal strain
ε_e	engineering strain (normal)
ε_{eu}	engineering strain at uniform elongation or necking
ε_f	true strain at fracture
ε_r	true strain in radial direction
ε_{max}	maximum absolute value of bending strain
ε_{tn}	true strain in the nth drawing pass
ε_t	true strain (normal)
ε_{tu}	true strain at uniform elongation or necking
ε_y	yield strain
ε_z	axial tensile strain
ε_θ	circumferential tensile strain
ε_o	effective strain
$\varepsilon_I, \varepsilon_{II}, \varepsilon_{III}$	principal strains
ϕ	angle of wire contact with capstan; angle of twist
ϕ_{sb}	angle of torsional springback
ϕ_y	angle of twist at initial yielding
Φ	redundant work factor
γ	shear strain
γ_f	shear strain at failure
γ_{max}	maximum value of shear strain
γ_y	shear strain at yielding
η	lubricant viscosity
η'	lubricant viscosity at ambient pressure
μ	coefficient of friction
μ_{max}	maximum coefficient of friction
ν	Poisson's ratio
θ_t	angle of twist
θ_{tmax}	angle of twist at failure
ρ	density
ρ_d	defect contribution to resistivity
ρ_e	electrical resistivity
ρ_i	impurity contribution to resistivity

ρ_T	temperature contribution to resistivity
σ	normal stress
σ_a	average flow stress
σ'_a	average flow stress in plane strain
σ_{an}	average flow stress in the nth drawing pass
σ_b	back stress on wire entering die
σ_d	drawing stress
σ_{do}	drawing stress in the absence of back stress
σ_e	engineering stress (normal)
σ_{eu}	engineering stress at uniform elongation or necking (UTS)
σ_g	single crystal flow stress
σ_m	maximum mean normal stress at the wire centerline during drawing
σ_{ns}	net section stress
σ_t	true stress (normal)
σ_{tu}	true stress at uniform elongation or necking
σ_y	yield strength
σ_o	flow stress, effective stress
σ_{To}	flow stress at reference temperature
σ_{00}	flow stress at the die entrance
σ_{01}	flow stress at the die exit
$\sigma_I, \sigma_{II}, \sigma_{III}$	principal stresses
$\sigma^\star$	maximum tensile stress
$\sigma^\star_{ave}$	average value of maximum tensile stress in a cross section
Σ	ratio of drawing stress to average flow stress
τ	shear stress
τ_y	shear stress at yielding
τ_{lub}	lubricant shear strength
τ_o	shear strength
ω	block, capstan or roll speed, in revolutions per unit time
Ω	ohms

REFERENCES

1. G. E. Dieter, *Mechanical Metallurgy*, Third Edition, McGraw-Hill, Boston, MA, 1986, 504.
2. B. Avitzur, *Handbook of Metal-Forming Processes*, John Wiley & Sons, New York, 1983, 195.
3. P. Körber and A. Eichinger, *Mitteilungen Kaiser Wilhelm Institute fur Eisenforschung*, 22 (1940) 57.
4. E. Siebel and R. Kobitzsch, *Stahl und Eisen*, 63(6) (1943) 110.
5. E. Siebel, *Stahl und Eisen*, 66–67(11–22) (1947) 171.
6. W. Linicus and G. Sachs, *Mitteilungen Material. Sonderheft*, 16 (1931) 38.
7. O. Hoffman and G. Sachs, *Introduction to the Theory of Plasticity for Engineers*, McGraw-Hill, Boston, MA, 1953, 176.
8. A. Pomp, *Stahldraht*, Second Edition, Verlag Stahleisen, Düsseldorf, 1952.
9. J. G. Wistreich, *Metallurgical Reviews*, 3 (1958) 97.
10. B. Avitzur, *Journal of Engineering for Industry Transactions ASME B*, 85 (1963) 89.
11. B. Avitzur, *Journal of Engineering for Industry Transactions ASME B*, 86 (1964) 305.
12. K. B. Lewis, *Steel Wire in America*, The Wire Association, Inc., Branford, CT, 1952.
13. R. M. Shemenski (Ed.), *Ferrous Wire Handbook*, The Wire Association International, Inc., Guilford, CT, 2008, 1.
14. J. E. Buchanan, *Wire: A Thread Through History*. In H. Pops (Ed.), *Nonferrous Wire Handbook, Vol. 3*, The Wire Association International, Inc., Guilford, CT, 1995, 1.
15. *The Making, Shaping and Treating of Steel*, 10th Edition, W. T. Lankford, Jr., N. L. Samways, R. F. Craven and H. E. McGannon (Eds.), Association of Iron and Steel Engineers, Pittsburgh, PA, 1985, 961.
16. S. Slesin, D. Rosensztroch, J. L. Menard, S. Cliff and G. De Chabaneix, *Everyday Things Wire*, Abbeville Press, New York, 1994.
17. J. N. Harris, *Mechanical Working of Metals*, Pergamon Press, New York, 1983, 208.
18. *CRC Handbook of Chemistry and Physics*, 90th Edition, CRC Press, Boca Raton, FL, 2009–2010.
19. J. G. Wistreich, *Proceedings of the Institution of Mechanical Engineers*, 169 (1955) 654.
20. R. N. Wright, *Wire Technology*, 4(5) (1976) 57.
21. W. A. Backofen, *Deformation Processing*, Addison-Wesley Publishing Company, Reading, MA, 1972, 139.
22. L. F. Coffin, Jr. and H. C. Rogers, *ASM Transactions Quarterly*, 60 (1967) 672.
23. H. C. Rogers and L. F., Coffin, Jr., *International Journal of Mechanical Science*, 13 (1971) 141.
24. R. Hill and S. J. Tupper, *Journal of the Iron and Steel Institute*, 159(4) (1948) 353.
25. R. N. Wright, Center Bursts — A Review of Criteria. In *2008 Conference Proceedings*, Wire Association International, Inc., Guilford, CT, 2008, 15.
26. S. Kalpakjian, *Mechanical Processing of Materials*, D. Van Nostrand Company, New York, 1967, 175.
27. R. N. Wright, *Wire Journal International*, 33(5) (1999) 118.
28. A. E. Ranger, *Journal of the Iron and Steel Institute*, 185 (1957) 383.
29. J. A. Schey, *Tribology in Metalworking, Friction, Lubrication and Wear*, American Society for Metals, Metals Park, OH, 1983, 135.
30. E. R. Booser, *Kirk-Othmer Encyclopedia of Chemical Technology*, 3rd Edition, Vol. 14, Wiley-Interscience, New York, 1981, 477.

31. O. Pawelski, W. Rasp and T. Hirouchi, *Tribologie*, Springer-Verlag, Berlin, 1981, 479.
32. R. N. Wright, *Wire Journal International*, 30(8) (1997) 88.
33. C. Noseda and R. N. Wright, *Wire Journal International*, 35(1) (2002) 74.
34. R. N. Wright, *Metallurgical Transactions*, 7A (1976) 1891.
35. R. N. Wright, *Wire Technology*, 6(3) (1978) 131.
36. R. N. Wright, *Wire Journal International*, 41(12) (2008) 62.
37. A. B. Dove, *Deformation in Cold Drawing and Its Effects, Steel Wire Handbook*, Vol. 2, The Wire Association, Inc., Branford, CT, 1969, 16.
38. R. N. Wright and A. T. Male, *ASME Transactions, Journal for Lubrication Engineering, Series F*, 97(1) (1975) 134.
39. R. N. Wright, Metal Progress, 114(3) (1978) 49.
40. S. M. Bloor, D. Dawson and B. Parsons, *Journal of Mechanical Engineering Science*, 12(3) (1970) 178.
41. D. A. Lucca and R. N. Wright, *ASME Transactions, Journal of Manufacturing Science and Engineering*, 118(4) (1996) 628.
42. R. N. Wright, *Wire Journal International*, 38(1) (2005) 76.
43. R. N. Wright, Characterization of Metalworked Surfaces. In *Metallography as a Quality Control Tool*, Plenum Publishing Company, New York, 1980, 101.
44. R. N. Wright, Friction, Lubrication and Surface Response in Wire Drawing. In *Metalforming Science and Practice*, J. G. Lenard (Ed.), Elsevier Science Ltd, Oxford, UK, 2002, 297.
45. G. Baker and R. N. Wright, Evaluation of Wire Surface Topography, *Nonferrous Wire Handbook*, Vol. 3, Wire Association International, Inc., Guilford, CT, 1995, 546.
46. R. N. Wright, *Wire Journal International*, 35(8) (2002) 86.
47. Horace Pops, Ft. Wayne, IN.
48. R. N. Wright, *Insulation/Circuits*, 20(13) (1974) 30.
49. A. T. Santhanam, P. Tierney and J. L. Hunt, Cemented Carbides, *Metals Handbook*, Vol. 2, 10th Edition, ASM International, Materials Park, OH, 1990, 950.
50. L. Corbin, Dies & Ceramics, *Nonferrous Wire Handbook*, Vol. 3, The Wire Association, Inc., Guilford, CT, 1995, 477.
51. G. Baker and R. N. Wright, *Wire Journal International*, 25(6) (1992) 67.
52. D. G. Christopherson, H. Naylor and J. Wells, *Journal of the Institute of Petroleum*, 40 (1954) 295.
53. J. G. Wistreich, *Wire and Wire Products*, 34 (1959) 1486, 1550.
54. R. M. Shemenski (Ed.), *Ferrous Wire Handbook*, The Wire Association International, Inc., Guilford, CT, 2008, 407, 418.
55. H. Ford and J. G. Wistreich, *Journal of the Institute of Metals*, 82 (1953–1954) 281.
56. I. L. Perlin and M. Z. Ermanok, *Theory of Wire Drawing, Metallurgia*, USSR, (1971).
57. T. A. Kircher and R. N. Wright, *Wire Journal International*, 16(12) (1983) 40.
58. P. B. Martine, Y. Yi and R. N. Wright, *Experimental Mechanics of Shaped Bar Drawing, Advanced Technology of Plasticity, Vol. II*, K. Lange (Ed.), Springer-Verlag, Berlin, 1987, 863.
59. Y. Yi, Metal Flow in the Drawing of Shaped Bar, M. S. Thesis, Rensselaer Polytechnic Institute,1981.
60. P. B. Martine, An Analysis of the Drawing of Hexagonal Bar from Round Stock, M. S. Thesis, Rensselaer Polytechnic Institute, 1982.
61. W. A. Backofen, *Deformation Processing*, Addison-Wesley Publishing Company, Reading, MA, 1972, 89.
62. A. F. Sperduti, Shaping and Flattening Non-Ferrous Wire, *Nonferrous Wire Handbook, Vol. 2*, The Wire Association, Inc., Guilford, CT, 1981, 319.
63. G. E. Dieter, *Mechanical Metallurgy*, Third Edition, McGraw-Hill, Boston, MA, 1986, 606.

64. Standard Test Methods for Tension Testing of Metallic Materials, *2009 Annual Book of Standards,* Vol. 03.01. ASTM International, Philadelphia, PA, 2009, 70.
65. Standard Test Methods for Tension Testing of Metallic Materials, 2009 *Annual Book of Standards, Vol. 03.01.* ASTM International, Philadelphia, PA, 2009, 64.
66. G. E. Dieter, *Mechanical Metallurgy,* Third Edition, McGraw-Hill, Boston, MA, 1986, 294.
67. S. Kalpakjian, *Mechanical Processing of Materials,* D. Van Nostrand Company, New York, 1967, 15.
68. J. F. Alder and V. A. Phillips, *Journal of the Institute of Metals,* 83 (1954–1955) 80.
69. A. T. Male and G. E. Dieter, *Hot Compression Testing, Workability Testing Techniques,* G. E. Dieter (Ed.), American Society for Metals, Metals Park, OH, 1984, 53.
70. *The Making, Shaping and Treating of Steel,* 10th Edition, W. T. Lankford, Jr., N. T. Samways, R. F. Craven and H. E. McGannon (Eds.), Association of Iron and Steel Engineers, Pittsburgh, PA, 1985, 1232.
71. C. Brady, National Bureau of Standards (now the National Institute of Standards and Technology, Gaithersburg, MD).
72. D. Hull, *Introduction to Dislocations,* Second Edition, Pergamon Press, Oxford, UK, 1975, 27.
73. S. Kalpakjian, *Mechanical Processing of Materials,* D. Van Nostrand Company, New York, 1967, 36.
74. S. Kalpakjian, *Mechanical Processing of Materials,* D. Van Nostrand Company, New York, 1967, 37.
75. H. W. Hayden, W. G. Moffatt and J. Wulff, *The Structure and Properties of Materials, Vol. III, Mechanical Behavior,* John Wiley & Sons, New York, 1965, 10.
76. H. W. Hayden, W. G. Moffatt and J. Wulff, *The Structure and Properties of Materials,* Vol. III, Mechanical Behavior, John Wiley & Sons, New York, 1965, 17-18.
77. R. N. Wright, Workability and Process Design in Extrusion and Wire Drawing, Chapter 21, *Handbook of Workability and Process Design,* G. E. Dieter, H. A. Kuhn and S. L. Semiatin (Eds.), ASM International, Materials Park, OH, 2003, 316.
78. G. E. Dieter, *Mechanical Metallurgy,* Third Edition, McGraw-Hill, Boston, MA, 1986, 262.
79. M. G. Cockcroft and D. J. Latham, *Journal of the Institute of Metals,* 96 (1968) 33.
80. P. W. Lee and H. A. Kuhn, *Metallurgical Transactions,* 4 (1973) 969.
81. R. N. Wright, *Wire Journal International,* 40(6) (2007) 49.
82. R. N. Wright, *Wire Journal International,* 38(4) (2005) 116.
83. R. N. Wright, *Wire Journal International,* 37(1) (2004) 56.
84. D. A. Metzler, Ultrafine Drawing of Copper Wire, *Nonferrous Wire Handbook,* Vol. 3, Wire Association International, Inc., Guilford, CT, 1995, 360.
85. G. J. Baker, Workpiece Wear Mechanisms in the Drawing of Copper Wire, Ph.D. Thesis, Rensselaer Polytechnic Institute, 1994.
86. G. Baker and H. Pops, Some New Concepts in Drawing Analysis of Copper Wire, *Metallurgy, Processing and Applications of Metal Wires,* H. G. Paris and D. K. Kim (Eds.), The Minerals, Metals and Materials Society, Warrendale, PA, 1996, 29.
87. A. K. Biswas and W. G. Davenport, *Extractive Metallurgy of Copper,* 3rd Edition, Pergamon, Elsevier Science Ltd., Oxford, UK, 1994.
88. Hazelett Strip-Casting Corporation, Colchester, VT.
89. F. Kraft, R. N. Wright and M. Jensen, Journal of Materials Engineering and Performance, 5(2) (1966) 213.
90. F. Kraft and R. N. Wright, An Improved Model for the Continuous Electrical Resistance Annealing of Copper Wire. In *Proceedings, 61st Annual Convention & 1991 Divisional Meetings,* Wire Association International, Inc., Guilford, CT, 1991, 28.
91. *Metals Reference Book,* C. D. Smithells (Ed.), Butterworths, London, 1976, 377.

92. W. D. Callister, Jr. and D. G. Rethwisch, *Fundamentals of Materials Science and Engineering*, John Wiley & Sons, New York, 2008, 259.

93. W. D. Callister, Jr. and D. G. Rethwisch, *Fundamentals of Materials Science and Engineering*, John Wiley & Sons, New York, 2008, 470.

94. D. T. Hawkins and R. Hultgren, Phase Diagrams of Binary Alloy Systems, *Metals Handbook, Vol. 8*, 8th Edition, American Society for Metals, Metals Park, OH, 1973, 295.

95. *Metals Reference Book*, C. D. Smithells (Ed.), Butterworths, London, 1976, 297.

96. *Metals Reference Book*, C. D. Smithells (Ed.), Butterworths, London, 1976, 489.

97. *The Making, Shaping and Treating of Steel*, 10th Edition, W. T. Lankford, Jr., N. T. Samways, R. F. Craven and H. E. McGannon (Eds.), Association of Iron and Steel Engineers, Pittsburgh, PA, 1985.

98. W. F. Smith, *Structure and Properties of Engineering Alloys*, Second Edition, McGraw-Hill, Inc., New York, 1993, 86.

99. W. F. Smith, *Structure and Properties of Engineering Alloys*, Second Edition, McGraw-Hill, Inc., New York, 1993, 89.

100. W. F. Smith, *Structure and Properties of Engineering Alloys*, Second Edition, McGraw-Hill, Inc., New York, 1993, 90.

101. C. D. Smithells (Ed.), *Metals Reference Book*, Butterworths, London, 1976, 510.

102. W. D. Callister, Jr. and D. G. Rethwisch, *Fundamentals of Materials Science and Engineering*, John Wiley & Sons, New York, 2008, 385.

103. W. D. Callister, Jr. and D. G. Rethwisch, *Fundamentals of Materials Science and Engineering*, John Wiley & Sons, New York, 2008, 387.

104. W. D. Callister, Jr. and D. G. Rethwisch, *Fundamentals of Materials Science and Engineering*, John Wiley & Sons, New York, 2008, 436.

105. W. D. Callister, Jr. and D. G. Rethwisch, *Fundamentals of Materials Science and Engineering*, John Wiley & Sons, New York, 2008, 432–433.

106. W. D. Callister, Jr. and D. G. Rethwisch, *Fundamentals of Materials Science and Engineering*, John Wiley & Sons, New York, 2008, 420.

107. W. D. Callister, Jr. and D. G. Rethwisch, *Fundamentals of Materials Science and Engineering*, John Wiley & Sons, New York, 2008, 429.

108. R. M. Shemenski (Ed.), *Ferrous Wire Handbook*, The Wire Association International, Inc., Guilford, CT, 2008, 83.

109. G. Lankford and M. Cohen, *Transactions of the American Society for Metals*, 62 (1969) 623.

110. R. N. Wright, *Wire Journal International*, 41(12) (2008) 62.

111. *The Making, Shaping and Treating of Steel*, 10th Edition, W. T. Lankford, Jr., N. T. Samways, R. F. Craven and H. E. McGannon (Eds.), Association of Iron and Steel Engineers, Pittsburgh, PA, 1985, 1284.

112. *The Making, Shaping and Treating of Steel*, 10th Edition, W. T. Lankford, Jr., N. T. Samways, R. F. Craven and H. E. McGannon (Eds.), Association of Iron and Steel Engineers, Pittsburgh, PA, 1985, 1400.

113. M. E. Donnelly, The Flow Stress of Low Carbon Steels under Hot Deformation: A Basic Data Compilation for Process Analysis, M. S. Project, Rensselaer Polytechnic Institute, 1982.

114. S. Mehta and S. K. Varma, *Journal of Materials Science*, 27 (1992) 3570.

115. P. S. Follansbee, J. C. Huang and G. T. Gray, *Acta Metallurgica et Materialia*, 38(7) (1990) 1241.

116. D. L. Pasquine, J. Gadbut, D. E. Wenschhof, R. B. Herschenroeder, C. R. Bird, D. L. Graver and W. M. Spear, Nickel and Nickel Alloys, *Metals Handbook Desk Edition*, American Society for Metals, Metals Park, OH, 1985, 15•22.

117. R. N. Wright, C. Baid and K. Baid, *Wire Journal International*, 39(3) (2006) 146.

118. *Electrical Wire Handbook*, J. K. Gillett and M. M. Suba (Eds.), The Wire Association International, Guilford, CT, 1983, 31.

119. P. N. Richardson, *Introduction to Extrusion*, Society of Plastics Engineers, Brookfield, CT, 1974, 4.

120. P. N. Richardson, *Introduction to Extrusion*, Society of Plastics Engineers, Brookfield, CT, 1974, 84.

121. R. M. Shemenski (Ed.), *Ferrous Wire Handbook*, The Wire Association International, Inc., Guilford, CT, 2008, 891.

122. A. R. Cook, Galvanized Steel, *Encyclopedia of Materials Science and Engineering*, Vol. 3, M. B. Bever (Ed.), Pergamon Press, Oxford, UK, 1986, 1899.

123. F. E. Goodwin and R. N. Wright, The Process Metallurgy of Zinc-Coated Steel Wire and Galfan® Bath Management, *Conference Proceedings, Wire & Cable Technical Symposium, 71st Annual Convention*, The Wire Association International, Inc., Guilford, CT, 2001, 135.

124. R. N. Wright and K. Patenaude, *Wire Journal International*, 34(5) (2001) 94.

125. R. R. Arnold and P. W. Whitton, *Metals Technology*, (3) (1975) 143.

126. R. N. Wright, Practical Analysis of Roll Pass Design, *Conference Proceedings, 70th Annual Convention*, The Wire Association International, Inc., Guilford, CT, 2000, 131.

127. R. Stewartson, *Metallurgical Reviews*, 4(16) (1959) 309.

128. Forming, Cold Heading and Cold Extrusion, R. N. Wright (Ed.), *Metals Handbook Desk Edition*, American Society for Metals, Metals Park, OH, 1985, 26•48.

129. A. T. Male and G. E. Dieter, *Hot Compression Testing, Workability Testing Techniques*, G. E. Dieter (Ed.), American Society for Metals, Metals Park, OH, 1984, 70.

130. W. A. Backofen, *Deformation Processing*, Addison-Wesley Publishing Company, Reading, MA, 1972, 148.

131. A. Tomlinson and J. D. Stringer, *Journal of the Iron and Steel Institute London*, 193 (1969) 157.

132. Forming, Forming of Bars, Tube and Wire, R. N. Wright (Ed.), *Metals Handbook Desk Edition*, American Society for Metals, Metals Park, OH, 1985, 26–35.

133. Forming, Cold Heading and Cold Extrusion, R. N. Wright (Ed.), *Metals Handbook Desk Edition*, American Society for Metals, Metals Park, OH, 1985, 26–50.

134. Top Products of 2008, *Wire & Cable Technology International*, 37(1) (2009) 30.

135. Top Products of 2008, *Wire & Cable Technology International*, 38(1) (2010) 42.

SELECTED FORMULAS

(See List of Symbols and equation number in text for more information)

Engineering strain	$\varepsilon_e = (\ell_1 - \ell_0)/\ell_0$	(4.1)
True strain	$\varepsilon_t = \ln(\ell_1/\ell_0) = \ln(A_0/A_1)$	(4.2)
Reduction	$r = 1 - (A_1/A_0)$	(4.6)
Percent reduction	$r = [1 - (A_1/A_0)] \times 100$	(4.7)
Delta, or deformation zone shape parameter	$\Delta \approx (\alpha/r)\,[1 + (1-r)^{\frac{1}{2}}]^2$	(4.9)
Drawing stress	$\sigma_d = w_u + w_r + w_f = \sigma_a\,[(3.2/\Delta) + 0.9](\alpha + \mu)$	(5.4, 5.13)
Redundant work factor	$\Phi \approx 0.8 + \Delta/(4.4)$	(5.8)
Draw stress to flow stress ratio	$\sigma_d/\sigma_a = \Sigma = [(3.2/\Delta) + 0.9](\alpha + \mu)$	(5.13)
Delta for draw stress minimization	$\Delta_{opt} = (1.89)\,(\mu/r)^{\frac{1}{2}}\,[1 + (1-r)^{\frac{1}{2}}]$	(5.16)
Die semi-angle for draw stress minimization	$\alpha_{opt} = (1.89)\,(\mu r)^{\frac{1}{2}}/[1 + (1-r)^{\frac{1}{2}}]$	(5.17)
Die pressure to flow stress ratio	$P/\sigma_a = \Delta/4 + 0.6$	(5.18)
Draw stress with back stress	$\sigma_d = \sigma_a\,[(3.2/\Delta) + 0.9](\alpha + \mu) + \sigma_b\,[1 - (\mu r/\alpha)(1-r)^{-1}]$	(5.20)
Equilibrated temperature after drawing pass	$T_{eq} \approx T_0 + \sigma_d/(C\rho)$	(6.1)
Distance beyond die for temperature equilibration	$L_{eq} \approx (vC\rho d^2)/(24K)$	(6.2)
Contribution of deformation to drawing temperature	$(T_w - T_0) = \Phi\,\sigma_a \ln[1/(1-r)]/(C\rho)$	(6.5)
Contribution of friction to drawing temperature	$(T_f - T_0) = \mu \cot\alpha\,\Phi\sigma_a \ln[1/(1-r)]/(C\rho)$	(6.6)
Maximum wire temperature at surface, in the die	$T_{max} = (1.25)\,\mu\,\Phi\,\sigma_a\,[(vL_d)/(C\rho K)]^{\frac{1}{2}} + \Phi\,\sigma_a \ln[1/(1-r)]/(C\rho) + T_0$	(6.7)
Coefficient of friction, from drawing parameters	$\mu = (\sigma_d/\sigma_a)\,[(3.2)/\Delta) + 0.9]^{-1} - \alpha$	(8.4)
Increase in wire diameter due to wear	$\delta = P\,(2\,L_{sliding}\,q/H)$	(9.2)
Length of wire drawn for a given die diameter increase	$L_{sliding} = (H\,\delta)/(2\,q\,P)$	(9.3)
Die life in time	$t_{life} = (H\,\delta)/(2\,v\,q\,P)$	(9.4)
Die life in mass drawn	$M = (\pi/4)\,L_{sliding}\,\rho\,d^2$	(9.5)

Percent slip	% slip $= 100 \ (V_{capstan} - V_1)/V_{capstan}$	(9.11)
Ratio of stress going onto capstan to stress coming off	$\sigma_d/\sigma_b = \exp(2\pi N\mu)$	(9.14)
Engineering stress	$\sigma_e = F/A_0$	(11.4)
True stress	$\sigma_t = F/A$	(11.6)
Percent area reduction at fracture	% area reduction $= [1 - (A_f/A_0)] \times 100$	(11.9)
Compressive stress, with friction	$P = \sigma_0 \ [1 + (\mu d)/(3h)] = \sigma_0 \ [1 + (\mu)/(3\Delta)]$	(11.13)
Buckling stress	$P = [(\pi d/(2h)]^2 \ (d\sigma/d\varepsilon) = [(\pi/(2\Delta)]^2 \ (d\sigma/d\varepsilon)$	(11.14)
Axial strain in wire subject to bending	$\varepsilon = (2y)/(D + d)$	(11.15)
Maximum axial strain in wire subject to bending	$\varepsilon_{max} = 1/(1 + D/d)$	(11.16)
Bending moment associated with yielding of wire	$M_y = [(\pi/(32)] \ d^3 \ \sigma_0$	(11.17)
Minimum ratio of mandrel to wire diameter, without fracture	$(D/d) = (1/\varepsilon_f) - 1$	(11.18)
Engineering shear strain in torsion	$\gamma = 2\pi \ y \ N_t/L$	(11.19)
Maximum shear strain in wire subject to torsion	$\gamma_{max} = \pi \ d \ N_t/L$	(11.22)
Twisting moment associated with yielding of wire	$T_y = (0.1134) \ d^3 \ \sigma_0$	(11.23)
Maximum number of twists possible, without fracture	$N_{tmax} = L \ \gamma_f/(\pi \ d)$	(11.24)
Springback formula for round wire	$1/R_1 = 1/R_0 - (3.4)(\sigma_0/E)(1/d)$	(11.25)
Annealing index	$I = Log_{10} \ (t) - Q/[(2.303)RT]$	(13.16)
Ratio of finishing width to starting width in bar rolling	$W_1/W_0 = n_s^2 \ (A_0/A_1)$	(17.2)
Bending radius associated with yielding	$R_y = d \ E/(2 \ \sigma_y)$	(18.4)
Limiting value of bending moment	$M_L = (0.147) \ d^3 \ \sigma_y$	(18.6)
Angle of twist associated with yielding of wire	$\phi_y = (2 \ L \ \tau_y)/(d \ G)$	(18.11)
Limiting value of torsional moment	$T_L = (\pi/12) \ d^3 \ \tau_y$	(18.13)
Torsional springback angle	$\phi_{sb} = (8 \ L \ \tau_y)/(3 \ d \ G)$	(18.14)
Extrusion pressure	$P = k \ \sigma_0 \ lnR = 2 \ \tau_0 \ [(mAsc/2A) + (\Phi \ lnR)]$	(18.21, 18.22)

INDEX

Printed in the United States
By Bookmasters